The Teaching of Science

The Teaching of Science

Wynne Harlen

David Fulton Publishers
London

David Fulton Publishers Ltd
2 Barbon Close, London WC1N 3JX

First published in Great Britain by
David Fulton Publishers 1992

Note: The right of Wynne Harlen to be identified as the author of this work has
been asserted by her in accordance with the Copyright, Designs and Patents Act
1988.

British Library Cataloguing in Publication Data

The teaching of science.
 1.
 I. Harlen, Wynne

 ISBN 1-85346-154-7

Typeset by Chapterhouse, The Cloisters, Formby, L37 3PX
Printed in Great Britain by BPCC Wheaton Ltd. Exeter

Contents

Foreword

The current low level of political and public understanding about science (and technology) will probably be viewed by our descendants as one of the backward features of contemporary Britain. Unfounded faith in the capacity of science and technology to continue to bail us out from the environmental and social troubles we create for ourselves co-exist with ignorant and arrogant dismissals of their potential for improving the human condition. A scientific approach to problem-solving is essential to progress on matters currently seen as beyond the traditional domains of science. What is central to the enterprise is the cultivation of a discipline in the collection and evaluation of appropriate evidence in the systematic testing of relationships between observations of phenomena in the world and the descriptions and explanations we offer for those observations. To date the specification of relevant principles and practices is best exemplified in science and technology, and one of the redeeming decisions in the chaos of changes introduced into schooling in the last twelve years has been the place given to science in the national curriculum, in both primary and secondary schools. The ways of thinking that are characteristic of scientists at work are best encouraged in young children, and Britain is fortunate that the protagonists for Primary Science have been so persistent and articulate. They have generated curricula which they have defended on the joint grounds of academic desirability and of feasibility in terms of children's competence. Wynne Harlen has been prominent among the advocates of Primary Science, and it is particularly for this series that she had found the time to write such an excellent text for it.

Peter Robinson
September 1991

Preface

Science was given its place in the primary curriculum rather later in the UK than in many other countries of the world. The urgency to make up the ground and to ensure that science is a continuous part of children's experiences in school from the age of five, has meant making changes in a hurry. There has not even been time for the initial teacher education courses to be developed before new teachers as well as those in practice have had to take in the implications of science as a 'core' subject. Teachers need a great deal of help. This book is a contribution, but it can be no more than that. It is written for teachers in schools and those in training. It is concerned with the planning, provision, assessment and evaluation of learning experiences for children and with the basis for decisions about all these, with the understanding of the nature of scientific activity, of children's learning and with the ideas, process skills and attitudes which are the objectives of their learning.

These are not the only kinds of help which are needed to further good practice of science in primary schools. For example, teachers also need more background information – it will be at least seventeen years before the first generation of teachers who have experienced science throughout their own schooling enter the profession. Equally, help is needed with in-school coordination of science, with communication to parents and others about the nature and aims of science education and with keeping up to date about relevant changes in educational policy and practice. These matters get a mention here, as they must in discussing classroom practice, but they are not treated in depth.

Having said what the book does not try to do I now need to say what I hope to have done. In general terms I have tried to put down the essential messages, as I see them, for the classroom teacher and to do this in a way which makes them easily accessible. The form of the book, in twenty six short chapters on well defined topics is intended to make for easy access and 'dipping' rather than continuous reading. Of course, it isn't easy to treat topics in isolation and there are numerous cross-references to other chapters where related information and discussion can be found. Accessibility in terms of communicating the messages is, I hope, helped by examples liberally sprinkled thoughout. I have drawn on many published sources for these as well as my own experience. All sources are, of course, acknowledged, but there are two in particular which I would like to mention here. The first is the journal *Primary Science Review*, a publication of the Association of Science Education, through which teachers and other practitioners share ideas and experiences of primary science teaching. I could not resist making liberal use of the fascinating examples which fitted so well as illustration of many points. The second is the materials published as *Match and Mismatch* in 1977, by the project which I directed from 1973 to 1977, with Sister Anne Darwin and Monica Murphy (now Hughes) as team members. These publications are now out of print and I have 'plundered' them here so that some valuable material is not entirely lost to teachers.

The national curriculum has loomed large in the sights of teachers in England and Wales and at first they did need specific help in absorbing it into their practice. That process is now well in hand and the national curriculum requirements are viewed in the perspective of their overall guiding role in ensuring similar learning opportunities for all children. We must not forget that the national curriculum in England and Wales is not the only curriculum statement in operation in the UK, let alone further afield. Using it is similar to using any national or regional curriculum framework. In this book the national curriculum is discussed and taken as the example in showing how a framework is used in planning, assessing, recording and reporting. But the book it not about teaching the national curriculum; it is about teaching science and teaching children.

I very much hope that readers will enjoy this book and find it helpful as a springboard to their own ideas. Its success will reside in the extent of its contribution to teachers' own enjoyment and understanding of practice and, through this, to their pupils' enjoyment and understanding of the world around them.

Wynne Harlen,
Edinburgh, 1991

CHAPTER 1

Science – What is it?

The question in this title may seem a tough one as a starter to a book for primary teachers. Some justification is called for, which is the first aim of this chapter. The main burden of the reason is that we teach according to how we understand the nature of what we are teaching and according to how we understand the nature of learning. The importance of the view of learning is taken up in the next chapter. Here we start with the understanding of scientific knowledge and scientific activity. Having justified its importance, the second aim is to describe a view of science which is consistent with present day thinking but may be different from the view received through our own science education.

The third aim is to distinguish between science and technology, between scientific activity and technological activity. This is not so as to suggest a separation of the two – technology has been an integral part of primary science for the past 25 years and there is no need to change this – but there are important differences of which we have to be aware if we are to foster the skills and understanding of both adequately.

The significance of the view of science

The question 'what is science' seems the most obvious and yet the most unnecessary question with which to begin. It is obvious that we should want to be clear about what it is we are teaching, not because we will necessarily be teaching directly about the nature of science but because, like it or not, we convey through our teaching a view of science and so it is necessary to have a 'feel' for what this is. Perhaps just as important is to have a feel for what is *not* science so that its characteristics are identified in both positive and negative terms.

If this does seem obvious then it is surprising and curious that many teachers avoid the subject of what science is, preferring not to spend time on 'all that philosophising'. The answer is perhaps that school science bears only a partial resemblance to what science is and so the nature of science becomes an academic point. Indeed school science, which has meant only secondary school science for many of us, has overemphasised the learning of individual facts. Facts which emerge from observations are indeed part of science but they are not the same as understanding, and teaching which goes little beyond these facts will lead to learning facts rather than learning science.

We need to change this because science is a major area of human mental and practical activity which generates knowledge, knowledge that can be the basis of important technological applications as well as of intellectual satisfaction. It is an important part of the education of all, not just of scientists, to be aware of the status and nature of scientific knowledge; how it is created and how dependable it is.

The teaching of science as facts is an illustration of the impact of the view of science on the learning experiences provided in school. Regarding scientific activity as being the application of principles and skills which first have to be learned leads to the aims of education in science being conceived as the teaching of these principles and skills. The dominant role of class activities is then seen as being to demonstrate the skills and to 'prove' the principles. Both explicitly and implicitly science is conveyed as

— objective
— capable of yielding ultimate truths
— 'proving' things
— having a defined and unique subject matter
— having unique methods
— being value-free.

If scientific activity is seen as developing understanding through testing ideas against evidence, with the ideas being accepted as being as good as the evidence which supports them, then the classroom or laboratory experiences will be rather different. Ideas will be explored rather than accepted and committed to memory and alternative views examined in terms of supporting evidence. These activities will portray science as

— a human endeavour to understand the physical world
— producing knowledge which is tentative, always subject to challenge by further evidence

— building upon, but not accepting uncritically, previous knowledge and understanding
— using a wide range of methods of enquiry
— a social enterprise whose conclusions are often subject to social acceptability
— constrained by values.

So how we view the nature of scientific activity is an important question. At the same time we don't have to go deeply into the philosophy of science to have a sufficient 'feel' for it. What is attempted here is a brief dip into the question of what is science which is hopefully enough to lay foundations for building a framework for teaching science. Like the foundations of any building, they will not subsequently be obvious, for it is not suggested that any of this is conveyed directly in teaching. Rather, again like other foundations, it should support all that comes later.

Views of what science is and is not

The social context of science

It is very easy to find statements about what science is not. John Ziman, one of those who has made significant contributions to the debate, has said that 'science is distinguished from other intellectual activity neither by a particular style or argument nor by a definable subject matter' (Ziman, 1968, p. 10). The truth of this is easy to argue. The subject matter of science is the biological and physical world around which can be equally the subject of history, art, geography or almost any discipline you care to mention. Similarly the methods of science, of observation, prediction, inference, investigation and so on, are used in the study and research of other subjects. From this simple argument it is apparent that science cannot be characterised as being either content or process alone.

Having said what it is not, Ziman goes on to give his idea of what science is, using the phrase 'public knowledge'. Less cryptically he means by this that science is a corporate activity. Scientists begin with the ideas of others, past and present. As Newton remarked, one can stand on the shoulders of giants and so see further. Each adds a contribution of his or her own ideas and passes the combined ideas to others, whose reactions determine their acceptability at that time. The history of science has many examples of ideas which have been 'before their time' and for which their proponents paid a price, although in a different social climate the ideas were later entertained more sympathetically. Given this view, it is not surprising that Ziman rejects a distinction between science as a body of knowledge, scientific activity and science as a social institution and seeks a unifying principle for science in all its aspects.

Science as falsifiable

Ziman's view begins from the assumption that facts and theories which are scientific have survived a period of critical study and testing by others which has persuaded the scientific community of their acceptability. He says nothing of the nature of the study and testing and appears to fall into the trap, as the philosopher Karl Popper would see it, of focussing on how scientific ideas are verified rather than how they are falsified. Popper's outstanding contribution to the debate about the difference between a scientific and a non-scientific theory is simply stated: a theory is scientific if it can be shown to be false by testing. This idea is not complicated. Often the greatest ideas have a simplicity which makes them seem obvious, once stated, and this is one of them.

A common sense way of explaining the importance of trying to show that a theory is wrong rather than that it is right is through thinking how easy it is to find support for an idea if you are looking just for this. It is easy to select evidence which is in agreement, to reinterpret that which does not agree and not to realise the existence, or to ignore the significance, of evidence which might challenge the idea. More important, however, is the situation where it is not that evidence is being ignored, but that it is conceptually impossible to find evidence which contradicts the idea.

To illustrate the point a good example is one used by Popper himself of psycho-analytical theories. He found that proponents of these theories, whether they were Adler's or Freud's, could always give an explanation of any human behaviour in terms of their chosen theory. Popper commented: 'I could not think of any human behaviour which could not be interpreted in terms of either theory. It was this fact – that they always fitted, that they were always confirmed – which in the eyes of their admirers constituted the strongest argument in favour of these theories. It began to dawn on me that this apparent strength was in fact their weakness' (Popper, 1988).

Popper contrasted this with Einstein's theory of general relativity, which is capable of being disproved by evidence. Einstein's theory was used to predict small changes in the apparent position of stars when light from them passes close to the sun. If found, these changes would support the theory but, if not, the evidence would refute it. Thus it was possible that evidence could be found either way. It appeared to Popper that here was the essential character of a scientific theory, that it could be tested and could be refuted by evidence. The psycho-analytical theories which could be used to explain every kind of human behaviour lacked this quality since any proposed test could, by definition, end only in a positive result.

Popper's idea about the essential character of a scientific theory has become part of a widely accepted view of science. Stephen Hawking, writing more recently, expressed it as follows:

> Any physical theory is always provisional, in the sense that it is only a hypothesis: you can never prove it. No matter how many times the results of experiments agree with some theory, you can never be sure that the next time the result will not contradict the theory. On the other hand, you can disprove a theory by finding even a single observation that disagreed with the predictions of the theory.
>
> (Hawking, 1988, p. 10)

Note that a theory does not have to be supported by the evidence to be scientific. Thus Newton's theory, which failed where Einstein's succeeded, remains scientific by the very fact that it was disproved. Scientific activity is characterised by developing theories which fit the evidence available but which may be disproved when further evidence comes to light, not by devising theories so malleable that they always fit the evidence, or for which there can be no disproving evidence (as in astrology).

So the ideas which children create can be scientific if they are testable and falsifiable and the fact that they are often falsified by the evidence makes them no less scientific. *Learning* science and *doing* science proceed in the same way.

Technology and science

Because science and technology have been intimately linked in the activities of primary school children, there often appears to be difficulty in distinguishing between them. There would certainly be difficulty in *disentangling* them, especially in relation to their role in practical activities where children are not only devising ways of problem solving and investigating but constructing actual devices to implement their ideas. But there should be no difficulty in *distinguishing* between science and technology, for they are quite different in aims and the kinds of activity through which their aims are pursued.

Scientific activity, as we have seen, aims at understanding. Technological activity uses

> knowledge and skills effectively, creatively and confidently in the solving of practical problems and the undertaking of tasks.
>
> (Layton, 1990, p. 11)

This statement sums up with remarkable clarity and economy of words the important aspects of technology. The main points to note are:

- The mention of 'knowledge and skills' not scientific knowledge and skills. Certainly scientific knowledge and skills are used very often in technology but they are not the only kinds; it is this which makes technology an aspect of the whole curriculum not an adjunct of science. In Layton's words, technology is 'a freshly-conceived, broad, balanced and progressive set of experiences designed to empower students in the field of practical capability and enable them to operate effectively and confidently in the made world.' (ibid)
- The reference to the effective, creative and confident use of the knowledge and skills. These qualities include not only aesthetic ability but the ability to find solutions to problems where compromises have to be made because of competing requirements, where resources are limited and where there are no existing guidelines to follow.
- The aims are described in terms of solving practical problems and undertaking tasks. This goes beyond the definitions of technology as solving problems relating to human need. Many tasks which are accomplished through technology (e.g. putting the letters in seaside rock) have little to do with human need but are to serve other purposes, often commercial competition, personal preference or national status.

Coming back to the classroom, we see some of these characteristics of technology in progress when children are building models, especially working ones, but in fact at all times when materials are used. There is some application of knowledge of materials, skill in fashioning them, compromise with the necessity of using the materials available and creativity in doing this to achieve the end result intended within the constraints of time and cost.

Technology is very closely interwoven with science at all levels. For example, advances in understanding of the universe depend on advances in the technology for receiving signals from outer space; progress in medical knowledge of the human body both depends upon and activates advances in instrument design. In the classroom, children working out how to test predictions based on their ideas often need to design and make something to suit their purpose. Their use of technology helps the development of their scientific understanding. At other times their technological activity may serve some quite different ends, as in craft or cooking or creating an obstacle course.

Distinguishing technological from scientific activities is important in teachers' minds because they are, as the above tries to show, different aspects of children's education. It makes sense, however, to continue to pursue both within the same topics and activities, just as these will also serve certain aims in mathematics, English and other subjects.

Children's work is multifaceted in the primary school. This book is about the scientific facet of this work but acknowledges that this exists in a mutually supportive role with other parts of the curriculum.

CHAPTER 2

Learning in Science

In Chapter 1 it was argued that the view of science which a teacher has influences his or her approach in the classroom. If science is regarded as a set of 'truths' and invariable procedures to be passed on, then there will be an emphasis on presenting these in a way which is regarded as best for promoting their understanding and acceptance. If science is regarded as the generation of ideas which make sense of observed facts, ideas which are tentatively accepted if they fit observations but which are always subject to the possibility of refutation, then the testing of ideas in relation to evidence is likely to play a much larger part. What you want to teach influences how you teach. The same point can be made about a teacher's view of how learning takes place and indeed may be of greater influence and significance because it will operate more widely than in one subject area.

The significance of a view of learning

These extracts from conversations with two teachers indicate how a view of learning implicitly lay behind their decisions about the way to organise their classrooms and to interact with the children. Both were teachers in junior schools in a large city; both schools were of similar, turn of the century architecture; both teachers taught classes of about thirty-two 10-11 year olds. The interviews were originally part of the *Match and Mismatch* materials (Harlen et al, 1977).

The first teacher arranged the desks in straight rows, all facing the front, where the teacher's desk and blackboard occupied prominent positions. She explained her reasons for this formality in arrangement which echoed the formality of her teaching as follows:

> Well, you see the main thing – I can't stand any noise. I don't allow

them to talk in the classroom . . . I just can't stand noise and I can't stand children walking about . . . in the classroom I like them sitting in their places where I can see them all. And I mean I teach from lesson to lesson. There's no children all doing different things at different times. . . . And of course I like it that way, I believe in it that way.

When asked why the desks were all facing the front, she replied:

. . . if they're all facing me, and I – well I can see them and I can see what they're doing. Because quite honestly I don't think that – when they're in groups, I mean all right they might be discussing their work. But I mean, how do you know?

As well as her own preference for quietness, she was convinced that it was from the blackboard and from herself that the children would learn. She worked hard to make her method work, setting work on the blackboard and marking books every evening. She did not think that children handling things or talking to each other had any role in learning.

The second teacher arranged the desks in blocks of four but in no particular pattern. There was no discernible 'front' or 'back' of the class. The children moved about freely and chatted during their work. She explained:

I hate to see children in rows and I hate to see them regimented. At the same time, you know, often I get annoyed when people think that absolute chaos reigns, because it doesn't. Every child knows exactly what they have to do . . . And, it's much more – you could say informal – but it's a much more friendly, less pressing way of working and . . . it's nice for them to be able to chat with a friend about what they're doing.

Asked if she worried that the chat may not be about work she replied:

Oh no. I mean obviously adults do when they work. As long as I get the end result that's sort of suited to that particular child I don't mind . . . If a child is not achieving what he should do then I do go mad because at ten and eleven they should know they've got a certain amount of work to do and the standard one expects of them.

She actively encouraged independence and considered that the class organisation helped in this:

I hate to think of children sitting – I'm not against formal education I'm not against informal, there are advantages to both methods – but I think the great danger of very formal teaching is that the teacher's seen as a tin god figure. And very often the children aren't given the opportunity to think for themselves.

The point here is not to suggest that either teacher is 'right' but that both have a clear rationale for the decisions they make about the organisation and methods they use in the classroom and that this is related to their views of how children learn. Both show consistency between what they do and what they want to achieve through it and so both are at ease in their classrooms.

It is worth noting in passing that to change the practice of either of these teachers would require far more than different teaching materials or insistence on rearranging the desks. There are masses of examples of teachers who have been asked to make organisational changes, which simply result in children sitting in groups but being taught as a class and having to twist their necks to see the blackboard. To make real changes requires a change in the teacher which is far more than adopting different ways of working; it means changing their views of how children learn and of their own role and that of materials in fostering learning.

A view of how children learn science

The following representation of learning is in terms of a model of what happens when someone encounters something new. As such it is particularly appropriate for describing the learning of children although it fits a good deal of adult learning as well. The model is itself an idea used to explain observations of an event, which in this case is how someone tries to make sense of a new event, object or relationship. It is based on evidence from observing children and adults making sense of new happenings and reflecting on what scientists do in trying to explain new observations.

When faced with something new we search around, often unconsciously, in our minds and use previous experience in trying to understand it. Suppose someone shows you a shell of a type, size and colour that you have never seen before. Although you could not identify it as the shell of a particular creature, it is likely that you would be able say that it had come from a living thing and was not man-made and perhaps even whether the creature was likely to have lived on land or in water. You would be linking previous experience to the observation of the texture, shape and density of the object. You may well have thought at first that it was made of china or

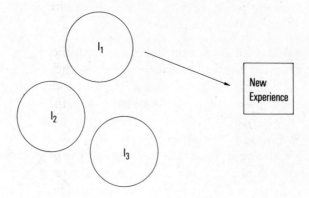

Figure 2.1

plastic but by tapping and fingering the object these ideas were dismissed, on the basis of knowledge from previous experience of the properties of these materials.

Children go through the same processes in trying to make sense of new objects or events, but because of their more limited experience they may not have an idea available to them which really fits and they use what seems most reasonable to them. For example, faced with the evidence that varnished cubes of wood stick to each other when wet several groups of eleven year olds concluded that the blocks became magnetic when wet (Harlen, 1985, p. 20). The resemblance of a block sticking to the underside of another, without anything to hold them together, to a magnet picking up an other magnet or a piece of iron was clearly very strong. An equally good alternative explanation was not available to them and so they held onto their view of magnetism, modifying it to accommodate the observation that the blocks only stuck together when wet by concluding that 'they're only magnetic when they're wet.'

These examples illustrate several points which are generally found in learning and which are represented in the model. Figure 2.1 is an attempt to show diagrammatically a number of existing ideas (I) which exist in the mind to be called upon to help understand the new experience. It may be that perceived similarities between a previous experience and the new one results in one or more of these ideas being linked in an attempt at understanding. Other processes also may create links – for example communication, since similar words used in description may suggest connections. It is not always logical reasoning and careful observation which leads to ideas being linked,

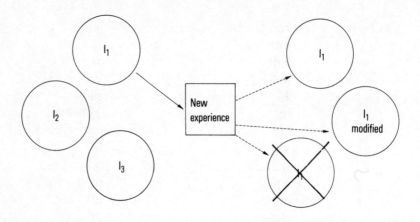

Figure 2.2

but creativity and imagination have a part. Indeed, in the case of the scientist faced with an unexpected phenomenon, it is the ability to try ideas from outside the immediately obvious that provides the start of a 'break through'.

But once an existing idea has been linked, its usefulness in *really* explaining the new experience has to be tested (Figure 2.2). The testing has to be done scientifically if the result is to be of value in making sense of experience.

In essence, this means that more evidence has to be sought to see whether it is consistent with predictions based on a linked idea or on possible alternative ideas. The 'testing' can be as simple as tapping the shell to see if it makes the kind of sound which would be expected if it were made of plastic. Here the idea that the object is made of plastic is a hypothesis which leads to a prediction and the tapping is the means of testing this prediction.

The nature of these linking and testing processes is discussed in Chapter 4. For the moment let us assume that the testing is carried out with scientific rigour and consider what may happen as a result. There are three main possibilities:

(i) that the linked idea is confirmed, in which case it becomes strengthened by successfully explaining a new phenomenon or having its range extended;
(ii) that the linked idea is not confirmed and so an alternative is tried (if there is one);
(iii) that the linked idea will fit the evidence if it is slightly modified.

Both (i) and (iii) result in a change in ideas which enables the new experience to be understood in terms of previous experience. Ideas have been extended or changed so that what was not previously understood can now be understood because it fits into ideas that make sense. In other words, learning has taken place.

This model, then, represents learning as a change in ideas, rather than the taking in of ideas from scratch. In practice, ideas which are not related to previous experience and thinking are of very little use to us. They do not arise from past experience and so are not called forth when past experience is used to explain new experience. Such ideas have to be committed to memory by rote and are usually usable only in contexts close to those in which they were learned. *Understanding*, which by definition is applicable, is created by development and change in ideas.

As we also saw in Chapter 1, there is much in common between learning science and doing science; indeed the process of advancing ideas at the frontiers of knowledge can be regarded as learning science in the way described. The scientist uses existing theories to attempt to explain new evidence and extends, rejects or adapts theories so that more phenomena can be understood and more problems solved.

Changing ideas into more scientific ones

The three main elements in this model of learning will be considered in more detail in the next three chapters. But there are two important points to make about the conditions under which ideas will become more scientific during the process of change embodied in the model.

First, when ideas are tested, the outcome in terms of changed or rejected ideas will depend on the way in which the testing is carried out. It was assumed in the argument above that the testing of ideas was rigorous and systematic, in the way associated with scientific investigation. When this is so, then ideas which do not fit the evidence will be rejected and those which do fit will be accepted and strengthened. But it may not be the case that the testing has this quality. The skills of young children – and those of some adults – have not developed to the appropriate degree. Children may ignore contradictory evidence in interpreting findings and hold on to their initial ideas even when these do not fit the evidence. Thus the extent to which ideas become more scientific (by fitting more phenomena) depends on the way in which the linking and testing are carried out,

14

that is, on the use of the process skills (see Chapter 4). *The development of understanding in science is thus dependent on the ability to carry out process skills in a scientific manner*. This is the reason for attention to development of these skills in science education; because of their role in the development of scientific concepts and not just because they are valuable skills in their own right.

The second point is that children, having less experience than adults, have fewer existing concepts to use in attempting to explain new events. It seems characteristic of human beings to try to explain things and that if ideas that really fit are not available then less satisfactory ideas will be used. It is more comfortable to modify an idea than to abandon it, especially if it is your only way of making any sense of an observation. It often happens that young children will hold onto their ideas to the point where their modification in order to explain away contrary evidence renders them unscientific (because untestable). Some examples show how this happens in practice.

Luis had an idea about what made snow melt, which was that it was caused by air rather than heat. He wanted to preserve some snow and said that it would not melt if it were put in jar with a lid on to keep out air. His first attempt led to the snow melting when the jar was brought into a warm room. He said that there was still some air there and that if the jar were to be packed with snow it would not melt. But however much snow was put into the jar he still said that there was room for air. He had, therefore, turned his claim into one which was unrefutable since by definition it would never be possible to have only snow in his jar.

Another example was a child who was convinced that something that did not float would do so if the water was deeper. More water was added but it was never enough; in spite of this, she maintained her claim that the object would float in very very deep water. Again the idea had become untestable.

In the interest of protecting the validity of hypothetical scientific ideas, it might be best, in these circumstances to agree the possibility of supporting evidence but to suggest that what had been seen so far did not seem to support the ideas. In such a situation it may be that the child will be prepared to test out another idea, which is put in terms of, 'well, let's try something else, perhaps . . .' but it may also be preferable simply to leave the matter until more experience has caused change in these ideas. This question of how to introduce alternative ideas is one which we shall pick up again in Chapter 10.

CHAPTER 3

Children's Ideas

The intention of this chapter is to substantiate the claim, which is at the heart of the view of learning described in the last chapter, that children form ideas about things around them long before they are 'taught' about them in school. The recognition that these ideas exist and the support given to the importance of taking them into account in teaching, adds urgency to finding valid ways of getting access to them. Examples from the Science Processes and Concepts Exploration project illustrate some techniques for doing this at the same time as revealing the nature of some of the ideas which many children have been found to hold.

Recognising the value of children's ideas

That some of children's ideas are different from those of adults has been known for a long time; it is nothing new. Before the second quarter of this century these ideas would have been dismissed as childish and silly; their expression perhaps tolerated as quaint in the home but certainly not in school. The significance of these ideas in the context of learning became recognised mainly through the work of Piaget. His publications from the early 1920s influenced many educators, such as Susan and Nathan Isaacs, who developed child-centred teaching approaches which acknowledged and valued children's ideas.

The spread of child-centred teaching approaches in primary science can be followed through the Nuffield Junior Science and Science 5/13 projects, both strongly reflecting Piaget's work, through to its reconstruction in the valuing of children's ideas by the Working Group which developed the national curriculum in science. The Working Group's recognition of the importance of children's ideas, expressed in its Final Report, was repeated in almost identical words in the Non-Statutory Guidance produced by the NCC:

> In their early experiences of the world, pupils develop ideas which enable them to make sense of the things that happen around them. They bring these informal ideas into the classroom and the aim of science education is to give pupils more explanatory power so that their ideas can become useful concepts. Viewed from this perspective, it is important that we should take a pupil's initial ideas seriously so as to ensure that any change or development of these ideas . . . becomes 'owned' by the pupil.
>
> (NCC 1989, A7 paragraph 6.2)

So the tenets of constructivist learning clearly now have widespread support, although it is necessary to look more closely at what taking 'a pupil's initial ideas seriously' means. The extent to which pupils' ideas are used, rather than merely revealed and noted, makes all the difference to whether or not the children have the intended 'owner-ship' as their ideas are modified. This is a matter which will be taken up in Chapter 6.

For the present, our focus in this chapter is to illustrate the nature of children's ideas, to give this phrase meaning in terms of actual ideas which children have been found to hold.

Ways of finding out children's ideas

Research work began on the scientific ideas of secondary pupils in the 1970s in New Zealand and then spread to Europe and North America (see, for example, the reader edited by Driver, Guesne and Tiberghien, 1985). The main work to date at the primary level has been carried out by the Science Processes and Concepts Exploration (SPACE), a joint project based at Liverpool University and King's College, London. Its research phase, which revealed the nature and extent of different ideas held by children across the whole range of concepts relevant at the primary level, was funded by the Nuffield Foundation. The methods for revealing children's ideas were of particular interest and have been built into the curriculum materials developed in the project's development phase.

Experience of working with children over many years placed some doubt over the applicability of methods used at the secondary level for ascertaining children's ideas. For example, much of the work of the Children's Learning in Science Project (CLIS, 1987) used test questions originating in the Assessment of Performance (APU) surveys. Pupils were given written questions, not necessarily connected with the science work they were doing at the time and asked to give answers on paper. Figure 3.1 shows two examples.

Other procedures used by researchers at the secondary level include interviews about either drawings depicting 'instances' of things happening which children were asked to explain or phenomena which took place at the time. Figure 3.2 is an example of an 'instance'.

It is not just the subject matter of these examples which seemed to make them less suitable for primary school children but two other features. First there is the difficulty of ensuring good communication, both through written words and through drawings, which may not recall to the child the same real events as those intended. Second is the possibility that the child may not have previously had the experience to think about the situation they are suddenly confronted with and asked to comment upon. In such circumstances the child may well think that the questions are silly ones and give answers which do not reflect serious thinking about the subject.

Methods developed by the SPACE project

These problems were avoided in the SPACE project by working in close cooperation with teachers. The teachers were asked to introduce into their classrooms materials and exploratory activities relating to the topics that the researchers wanted to discuss with the children. The teachers were asked not to provide any direct teaching about the materials but to encourage the children to observe and interact with the materials, which might be displayed on a table at the side of the classroom, with questions on cards inviting children to explore them. For example, when children's ideas about sound were to be discussed, the materials provided were a range of musical instruments, home-made and conventional, and other articles which could be used to make non-musical sounds. When interest was in ideas about growth the materials were growing plants, seeds which were germinating, stick insects and even an incubator with hens' eggs hatching.

For a period of about three or four weeks, teachers were asked to engage their children in discussion of these materials, to encourage them to keep records, in the form of a diary, of things which changed and to make drawings which tried to explain what was happening.

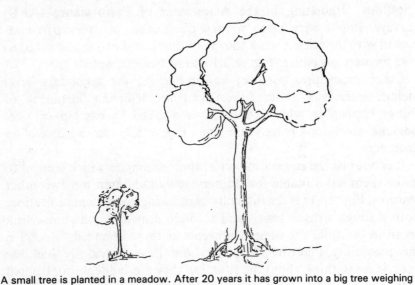

A small tree is planted in a meadow. After 20 years it has grown into a big tree weighing 250 kg more than when it was planted.

Where does the extra 250 kg come from? Explain your answer as fully as you can.

....

....

....

....

(Brook, Briggs and Driver, 1984)

After many experiments, scientists now think that:

- all things are made of small particles
- these particles move in all directions
- temperature affects the speed they move
- they exert forces on each other.

Use any of these ideas to help answer the following questions:

A football is pumped up hard during the day when it is warm. In the evening when the temperature falls, the football does not feel so hard. How does this happen; (Assume the football does not leak.)

....

....

....

....

(Bell and Brook, 1984)

Figure 3.1

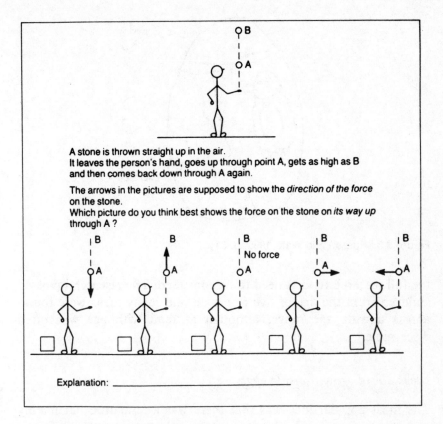

A stone is thrown straight up in the air.
It leaves the person's hand, goes up through point A, gets as high as B
and then comes back down through A again.

The arrows in the pictures are supposed to show the *direction of the force*
on the stone.
Which picture do you think best shows the force on the stone on *its way up*
through A ?

Explanation: _____

Figure 3.2 (Gunstone and Watts, 1985, p. 94)

The purpose of this activity was to give children the chance of experiences to think about and to encourage them to express their ideas. In addition, at the end of this period, a sample of children from each class was interviewed individually by the research team. The interview involved discussion of the same materials and activities in which the children had been engaged intermittently throughout the previous weeks. In this way many of the anticipated problems – of suddenly asking children questions about which they may not have been thinking – were avoided. (The research methods are described and illustrated in full in the SPACE Research Reports,1990, 1991).

Examples of children's ideas

The following examples represent some of the most common of children's ideas but they cannot represent the full variety. For that

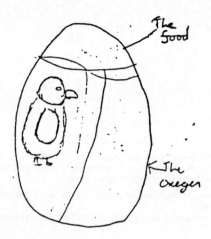

Figure 3.3 (Russell and Watt, 1990, p. 31)

the full research reports need to be consulted. The research involved children from the age of five to eleven and many ideas were found across a wide age range, although a trend with age was often discernible.

Ideas about growth inside eggs

The most popular idea was that there was a miniature but mainly complete animal inside the egg, feeding on what was there. This is evident in the drawings made by the children when asked to depict what they thought was inside an egg whilst it was incubating. (Figure 3.3). An alternative was that the complete animal was inside simply waiting to hatch, (Figure 3.4). There was also the view that the body parts were complete but needed to come together, (Figure 3.5).

The more scientific view that transformation was going on was evident in some children's ideas and it was also clear that they used knowledge derived from experience of reproduction of pets and observations of human babies in trying to understand what was going on inside the eggs.

Ideas about growth in plants

Infant children generally mentioned one external factor in response to the question 'What do you think plants need to help them grow?' For example, Figure 3.6 suggests that light is necessary.

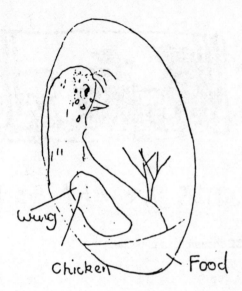

Figure 3.4 (ibid, p. 10)

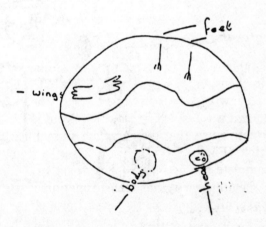

Figure 3.5 (ibid, p. 26)

Other young children mentioned soil or water or sun, but rarely all three. Characteristically the younger children made no attempt to explain why these conditions were needed or by what mechanism they worked. Junior children, however, made efforts to give explanations, as in Figure 3.7

The Plants grow by The window

Figure 3.6 (SPACE research, unpublished)

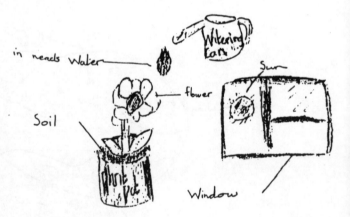

in needs Water

flower

Soil

Sun

plant pot

Window

I think that the plant
needs. water to ceap it alive
I also think that it needs some soile
to to help the plant to grow. The plant
sucks the soil up the stom. the soil
has got smelling in it to help it grow.
The plant needs sun because if it
did not have light it would never open its
petales.

Figure 3.7 (SPACE research, unpublished)

Figure 3.8 (Watt and Russell, 1980, p. 36)

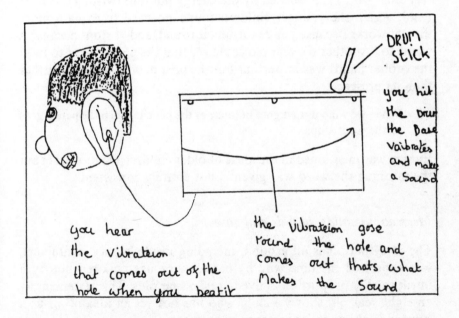

Figure 3.9 (SPACE research, unpublished)

Ideas about how sounds are made and heard

The example in Figure 3.8 suggests no mechanism for sound being produced by a drum or for it being heard; it is as if being 'very loud' and 'listening hard' are properties which require no explanation.

The simplest mechanism suggested is that the impact of hitting produces 'sound'. In contrast, Figure 3.9 explains the sound in terms of vibration. But notice that the vibration comes out of the drum through 'the hole'. A very common understanding of children was that sound travelled through air, or at least through holes in solid objects and not through the solid itself.

The notion of 'vibration' was associated with sound in ambiguous ways, sound sometimes being the same as vibration and sometimes having some cause and effect relationship to it. Figure 3.10 illustrates this struggle to connect the two.

Ideas about forces

Children's ideas about how things are made to move and what makes them stop were explored in various contexts including the 'cotton reel tank' which is propelled by the energy put into twisting a rubber band. Again the younger children found no need to explain more than 'it works because you're turning it round' and 'it stops because it wants to.' Another six year old could see that the pencil (used to twist the rubber band) was important but the idea of why went no further than its presence:

> When we wind it up it goes because of the pencil. When the pencil goes to the tip it stops.

Energy was mentioned in the ideas of older children (Figure 3.11) but the meaning the word was given is not entirely consistent.

Ideas about solids, liquids and gases

The idea that air is all around, including inside 'empty' containers, was expressed in some way by most junior age children but by a much smaller proportion of five to seven year olds. This statement by an eight year old shows a child who has not yet an idea of air as a substance although its presence is accepted:

> You can't see the air, but sometimes you think there is nothing in there 'cos you can't see anything, but it isn't a matter of seeing, it's a matter of knowing.

Even young children have relatively little difficulty in identifying something hard, such as steel, as solid, and something watery as a

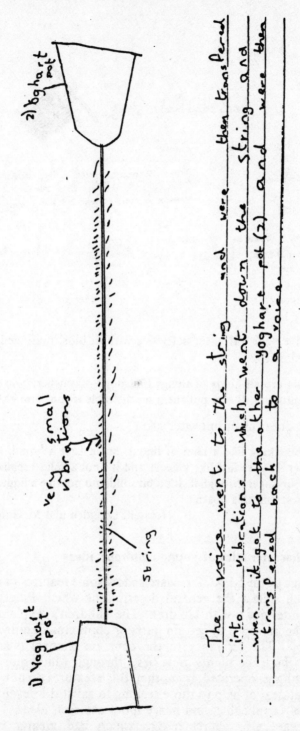

Figure 3.10 (SPACE project, unpublished)

The voice went to the string and were then transfered into vibrations which went down the string and when it got to the other yoghart pot (2) and were then transfered back to a voice.

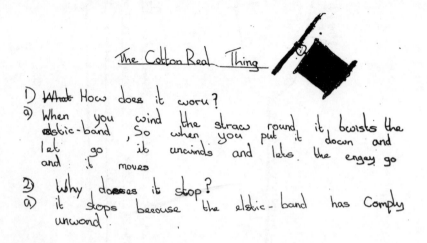

The Cotton Reel Thing

1) What How does it work?

2) When you wind the straw round, it twists the elastic-band, So when you put it down and let go it unwinds and lets the engay go and it moves

2) Why doesses it stop?

a) it stops because the elstic-band has Comply unwond.

Figure 3.11

liquid. A five year old, after activities with liquids, managed to give a general definition:

> Liquids are all kinds of things that don't stay where you put them. They just run. If you put them on the table it runs along the table.

But where does talcum powder fit?

> It's something like a kind of liquid, but it isn't a liquid, it's talcum powder. It goes fast like vinegar and it's not a solid because you can put your finger through it. It's a bit solid, but not like a liquid. A liquid feels wet and this doesn't.

> (Russell, Longden and McGuigan, 1991)

General characteristics of young children's ideas

The findings of the SPACE research show some features of children's ideas which reflect the general descriptions which Piaget distilled from his interviews with children. The children's 'explanations' in terms of the presence of certain parts or contextual features provide an example. The reference to the concrete channel (such as 'the window', where the plants grew best) through which more abstract causes (sunlight) operated, is another. Such features do not, however, give a great deal of help to those seeking to assist the development of more widely applicable, and hence more abstract, ideas.

There were also characteristics which had greater immediate relevance to teaching. In some case these are only hinted at in the examples given above, but they have a considerable basis of evidence in the full research findings. The children's ideas:

(i) emerged from a process of reasoning about experience, rather than from childish fantasy or imagination;

(ii) would not, however, stand up to rigorous testing against evidence that was often available for the children to use had they wished to do so;

(iii) sometimes required additional evidence to be made available if they were to be tested in practice;

(iv) were influenced by other information than that which came from evidence of actual events, these other sources including the media, conventions of speech and of ways of representing things, influential adults and peers;

(v) were often expressed in terms of words which seemed scientific, yet had, for the children, a meaning which was ill-defined, difficult to pin down and not apparently consistent with the scientific meaning.

These features of children's ideas are important in relation to 'taking their ideas seriously' and will be revisited in that context in Chapter 6.

CHAPTER 4

Process skills used in science

The important role of process skills in the development of children's ideas has been discussed in Chapter 2, although without specifying the range and nature of the process skills. It is the purpose of this chapter to consider how to identify the process skills and to describe them in action. We shall return to the matter of progression in developing the skills and consider the role of the teacher in this development in Chapter 11.

Process skills are described in various ways, all of which suffer from the problem of trying to draw boundaries round things which are not separable from each other. This will soon be apparent here, for when we describe an example of 'observing' there is some 'hypothesising' going on as well and even some degree of 'investigating'. Yet more obvious is that almost any scientific activity begins with 'observation' and so it is an integral part of other process skills. In the light of these points it is reasonable to ask, how useful is it to attempt to separate aspects of scientific activity? It may be best to regard it as a whole. However, the whole is so complex that, whilst admitting that they are not separable in practice, it is useful to describe certain aspects of scientific activity and to name them. In this way we can at least hope to arrive at a common and clear notion of the parts which are interwoven in the investigation of the world around and the development of understanding. It will soon become apparent that these aspects of practice which we call process skills, are not single skills but conglomerates of coherent skills. It is for convenience only that we refer to each as individual skills.

 There are overlapping sets of process skills involved in linking and testing ideas. Those involved in linking existing ideas to try to understand new experience can be listed as: observing, hypothesising, predicting and communicating; those involved in testing ideas are:

predicting, planning and carrying out investigations, observing, interpreting findings, drawing conclusions from them and communicating. Putting these together, the list of skills that will be considered is:

Observing
Hypothesising
Predicting
Investigating
Interpreting findings and drawing conclusions
Communicating

The description of each one is mainly in terms of a list of actions which anyone involved in using the process skill may be carrying out. The items in each list are not in any particular order at this stage.

Observing

For an image of observing in action let's return to the example of someone holding in their hand an object which resembles a shell; it might be a shell of a kind they have never seen before – or it might not be a natural object.

They rub their fingers over it, hold it to the light to see the detail (or use a hand lens if there is one), they smell it, even put it to their ear (as people do with shells); they might tap it and listen to the sound. They may mentally, or actually, be comparing it with something known to be a shell, seeking points of similarity and difference. All these things add up to careful observation in this particular situation.

In another example, imagine a person watching a 'cartesian diver', made from a dropper floating in water inside a large plastic bottle. When the bottle is standing, firmly capped, the dropper floats at the surface upright, with water about half way up the tube. When the sides of the bottle are squeezed the dropper slowly sinks to the bottom. More than this can be noticed, however, by looking carefully and noting the order in which events take place. The level of the water inside the dropper's tube rises when the bottle is squeezed and this happens before the dropper starts to sink; indeed it could be the reason for the

sinking (which would be a hypothesis based on the observation). At this point, however, we keep the focus on the observation. In this situation the careful observation of detail and of the order of events are important aspects of what we mean by observation in action.

From thinking about these and other examples we can identify the kind of actions which indicate observation in action as including the following:

- making use of several senses
- noticing relevant details of the object and its surroundings
- identifying similarities and differences
- discerning the order in which events take place
- using aids to the senses for study of details.

Hypothesising

In the cartesian diver example a hypothesis was mentioned in the context of attempting to explain why the dropper sank. A hypothesis is a statement put forward to attempt to explain some happening or feature. When hypothesising the suggested explanation need not be correct, but it should be reasonable in terms of the evidence available and possible in terms of scientific concepts or principles. There always is some knowledge brought to bear on the evidence in making a hypothesis. In the case of the dropper, the suggestion that it sank because of the extra water inside it was based not only on the observation of the water level in the tube but also the knowledge that adding more mass to a floating object can make it sink.

Often there is more than one possible explanation of an event. This underlines the point that hypotheses are plausible but not necessarily correct. Take for example the observation that one of two puddles of water left after a rain storm dries up more slowly than the other. The reason could be that there was more in one than the other to begin with, that one is on more water permeable ground than the other, that only one is in the sun, that there is greater air movement over one than the other. Several more might well be thought up. In each case there is knowledge brought to bear - about the conditions which favour evaporation, about the differences in properties of materials with regard to water permeability, and so on. If the suggestion were that the water ran uphill out of one of them or that it dried up more quickly 'because it wanted to', these would not be scientific hypotheses because they conflict either with the evidence or with the scientific knowledge.

Even though one may not oneself be able to think of alternative explanations it is important to recognise the possibility of alternatives. In turn this brings the realisation that any hypothesis has to be regarded as provisional since there may always be another which is more consistent with the evidence.

So the indicators of hypothesising include:

- suggesting an explanation which is consistent with the evidence
- suggesting an explanation which is consistent with some scientific principle or concept
- applying previous knowledge in attempting an explanation
- realising that there can be more than one possible explanation of an event or phenomenon
- realising the tentative nature of any explanation.

Predicting

A prediction is a statement about what may happen in the future, or what will be found that has not so far been found, that is based on some hypothesis or previous knowledge. For example, if you know how far a car will go on two gallons of petrol and how far it will go on four gallons, it is possible to predict how far it will go, say, on five gallons. In another cases, if your hypothesis about why a table lamp is not working is that the fuse has blown, then you predict that changing the fuse will make it work again.

A prediction is quite different from a guess, which cannot be justified in terms of a hypothesis or evidence. Given no information about how far the car goes on a certain amount of petrol the suggestion of how far it would go on five gallons would be a guess, not a prediction (except that evidence of other cars may be used, in which case the prediction would be based on that evidence). However even when there is a rational basis for a prediction there is always the possibility that it will not be supported by evidence of finding out what actually does happen. This may be because a relationship assumed to hold does not do so indefinitely. For example, an elastic band may stretch 10cm under a force of 5 Newton and 20cm under a force of 10 Newton but break under 15 Newton! Caution is thus needed in making a prediction which depends on applying a relation-ship beyond the range of available evidence, (extrapolating). Interpolating (predicting within the range of evidence), such as predicting how far the elastic band will stretch with a force of 8 Newton is much safer.

There is often confusion of the meaning of a hypothesis and a prediction, partly because the hypothesis on which a prediction is based may be implicit, not explicit. So, a statement such as 'I will be able to see myself better in that spoon than this one because it is shinier' is a prediction The related hypothesis is that shininess makes surfaces reflect better.

The behaviours which indicate that predicting is in action thus include:

• making use of evidence from past or present experience in stating what may happen
• explicitly using pattern in evidence to extrapolate or interpolate
• justifying a statement about what will happen or be found in terms of present evidence or past experience
• showing caution in making assumptions about a pattern applying beyond the range of evidence
• distinguishing a prediction from a guess.

Investigating

The whole range of process skills could be regarded as being part of 'investigating' and some people do in fact regard it as such and therefore omit it from a list of constituent skills. However, a slightly more restricted meaning can be given, covering what happens from the point where a question for investigation has been raised, or a prediction made on the basis of a hypothesis which has to be tested, to the point where evidence is gathered and needs to be interpreted. Even this more restricted meaning, which is taken here, involves bringing together several kinds of thinking and action concerned with planning and carrying out investigations. Sometimes it is helpful to separate planning from the practice of carrying out an investigation. But often the two occur together, particularly in the case of children, and parts of the investigation are worked out along the way rather than being all thought through before taking action.

As an example of what investigation involves, suppose you do decide to investigate possible reasons for one puddle of water drying up more than the other. The first hypothesis which you decide to test is that it could be caused by different movement of air over the water.

But there are a number of things which can vary (variables) and to find out whether air movement can make a difference, it is necessary to set up a test where this is the only thing which is different. It is necessary to create experimental conditions such that other things can be kept the same. The 'puddles' need to have the same amount of water, be on the same kind of ground, receive the same sunshine and be the same in any other respect which is thought might affect their rate of drying. These considerations refer to the variables which have to be kept the same, or controlled. It would be much easier to ensure that 'all other things are equal' by making 'puddles' in equal containers indoors. One variable, the amount of movement in the air, has to be different, of course, in the two cases. This is the independent variable, the one which is being changed by choice. An investigation set up in this way is described as 'fair' in the sense that there is (or should be) no other variable than the one we choose to change which will affect the result and so that result can be said to have been caused by the independent variable.

What may change as a result of the change in the independent variable, in this case the amount of water left, is dependent on what is done and so is called the dependent variable. Any resulting change in this variable has to be compared or measured to obtain the result and this has to be done in a suitable way. Thus first planning ways of doing this, using measuring instruments if necessary, and then using the equipment effectively are all parts of the process of investigating. If the result shows that there is no difference resulting from varying the air movement, then another investigation may be carried out with another variable, say the exposure to sun, as the independent variable. In that investigation the air movement would be kept the same and not varied.

Indicators of this skill in action would include all these actions involved in setting up and carrying out an investigation:

- deciding which variable is to be changed (independent) and which are to be kept the same (controlled)
- carrying out the manipulation of variable so that the investigation is 'fair'
- identifying which variable is to be measured or compared (dependent variable)
- making measurements or comparisons of the dependent variable using appropriate instruments
- working with an appropriate degree of precision.

Interpreting findings and drawing conclusions

Interpreting involves putting results together so that patterns or relationships between them can be seen. The results of water evaporation investigation might be in terms of the amounts of water present in two containers at various times. To interpret these would mean relating what happened to the different conditions in a statement such as 'the water went down more quickly when the air was moving than when it was still'. The further step of drawing a general conclusion, such as 'water evaporates more quickly in moving air than in still air' requires some caution, since it suggests not only that this relationship was found in one particular investigation but that there is reason to suppose that it would hold in other cases.

Where several pieces of information have been collected (as in seeing how far an elastic band stretches under various forces), interpretation involves looking for patterns in them. These patterns might be regular – as in the case of the elastic band, as long as it is not stretched too much - or merely trends – such as in the tendency for taller people to have larger hands and feet than shorter people. In the regular pattern case all the information will fit the pattern without exception, but in the latter there is an overall association although there may be exceptions to it in some cases.

If there are only two sets of data, as in the comparison of the rate of evaporation of the two puddles, it is not justified to claim a pattern or trend. The interpretation would only be justified in terms of what happens with and without movement of air. Only if there were at least three different conditions (e.g. still air, slowly moving air, fast moving air) would it be possible to draw a conclusion about the effect of wind speed on the rate of drying.

It is often tempting to jump to the conclusion that a pattern exists on the basis of only some of the data, ignoring other information which conflicts. Thus checking a pattern or trend against all available information is important. Even then the possibility that more data, were it to be collected, might conflict with the pattern has to be kept in mind. 'On the basis of our results . . .' would be a fitting start to a statement of a pattern or conclusion. Thus indicators of this skill in use are:

- putting various pieces of information together to make some statement of their combined meaning
- finding patterns or trends in observations or results of investigations

- identifying an association between one variable and another
- making sure that a pattern or association is checked against all the data
- showing caution in making assumptions about the general applicability of a conclusion.

Communicating

Talking, writing, drawing or representing things in other ways are not only means of letting others know of our ideas but help us to sort out what we think and understand. Thus communication is important in learning and takes various forms according to the subject being learned. In science, talking and listening are particularly valuable for making ideas explicit and for helping the understanding of scientific vocabulary. Children often use words which they pick up without necessarily realising the meaning which attaches to the words. Encouraging children to talk about what they mean by these words and to listen to what others say can help to reveal differences which can be the source of misunderstanding. Discussion also helps where children may have ideas but no words to put to them. There is more about the use of scientific words in Chapter 13.

Reports children write at the end of an investigation to present what they have found are only one form of writing, one which is most useful when it is seen to have a specific purpose and a specific audience. Notes or drawings made for themselves during the course of an investigation are valuable for jotting down ideas as well as findings and other information. They can take the same role as talking with regard to helping to sort out ideas, as pointed out later (Chapter 11, p. 92)

Communication in science involves using various conventions of representation which help in organising information and conveying it efficiently. Graphs, charts, tables, symbols, etc serve this purpose and have to be chosen to suit the particular kind of information. Communication is, of course, two way, and involves the ability to take information from written sources, to use information presented in graphical or tabular form, thus expanding the evidence which can be used in testing ideas.

Thus indicators of scientific communication skills include:

- talking, listening or writing to sort out ideas and clarify meaning
- making notes of observations in the course of an investigation

- using graphs, charts and tables to convey information
- choosing an appropriate means of communication so that it is understandable to others
- using secondary sources of information.

CHAPTER 5

Attitudes related to learning science

The concern in this chapter is the general disposition to act or react in certain ways which we describe as 'attitudes'. The importance in learning of these aspects of behaviour is succinctly captured in these words by K. M. Evans:

> Ask a boy what he learns at school and he will tell you English and Mathematics, History and Geography, Science and Languages. But his teachers, if not the boy himself, will know that he learns far more than this. Models of thinking and acting, attitudes and interests are also acquired and developed during these school days, and these may become permanent, remaining effective and observable long after the greater part of the subject matter learnt has been overlaid or forgotten. It would be difficult to overstress the influence of attitudes and interests in the lives of individual people. They determine what a man will do or say in particular situations, what he will enjoy or dislike, his approach to other people, and his reactions to events in his own life and in the world around.
>
> (Evans, 1965, p. 1)

As this passage suggests, there are many different kinds of attitudes which influence behaviour. Our concern is with those which influence children's learning and in particular their learning in science at an early stage. There are at least four kinds of attitudes which are relevant:

— attitudes towards school work
— attitudes towards themselves as learners
— attitudes towards science as an enterprise
— attitudes towards objects and events in the environment.

We shall consider these in two pairs: general attitudes towards the self in relation to school work and scientific attitudes. For each pair the concern is with why the attitudes are important and how they develop. What teachers can do to encourage positive attitudes is taken up in Chapter 12.

Attitudes towards self in relation to school work

Importance

For all of us, what we can do is influenced by what we feel about it and what we feel we can do. So with children; those who feel they can succeed are more likely to do so, whilst those who anticipate failure are less likely to succeed. These attitudes affect not only what is learned but the effort put into the tasks given, which in turn affects the likelihood of success. When the attitudes are negative there is a vicious circle created:

— the child thinks he or she is no good at the task
— the child thinks it is not worth making an effort
— little success results from the lack of effort
— the feeling of being 'no good' is reinforced.

Feelings about school in general and about different activities are thus very important in developing the self-confidence needed to tackle new tasks. So how do these feelings develop?

Development

For the young child, ideas and feelings about the self are formed very largely from the ways in which he or she is treated by others. The picture the child has of him/herself grows from comments and actions of others around him or her, at home at school and in social groups. These things reflect the image others have of the child and have a strong influence on what (s)he thinks (s)he is. Parents often try to protect their children from failure by lowering expectations: 'You're not very good at drawing, are you, Jane, so you'd better leave it to Meg' or talk about them to others in their hearing: 'Ali's like me, not very bright!' It is easy to see how these images which others have of children can become their self-images: they will see themselves as 'no good' at drawing, or whatever, and expect to be grouped among the 'not very bright'.

In school, teachers and other children make comments every day which reinforce the general trend of children's performance. The classroom organisation in which children are grouped by ability on different tables leads to children labelling themselves, however the teacher tries to use neutral group names: 'We're the green table; we won't be able to do the same as the others'. Work becomes something to endure for such children, who then often attempt to preserve or boost their self-esteem in other ways; sometimes this can be positive but it also takes the form of 'rubbishing' others in order to show themselves in a better light.

Of course children do vary in achievement. The challenge is to cater for this without labelling them, to build up self-confidence so that there is always willingness to try and thus have the chance of success. When this happens the vicious circle can be turned into a virtuous circle, from which a positive self-image emerges; realising that one's achievement is less than others' can be accepted without bringing with it a sense of failure as a person.

Scientific attitudes

The distinction was made between two kinds of attitude:

— those *towards science as an enterprise*
— and those *towards objects and events in the environment which are studied in the course of scientific activities.*

To form an attitude towards science it is necessary to have an idea of what 'science' is.

There are many myths about science and about scientists which persist in popular belief and the caricatures which are perpetuated in the media and in some literature. Typically these portray scientists as male, bespectacled, absent minded and narrowly concerned with nothing but their work. Science as a subject may be portrayed as the villain, the origin of devastating weapons and technology which causes environmental damage or as the wonder of the modern world in providing medical advances, expanding human horizons beyond the Earth and responsible for the discoveries which lead to computers and information technology. We should avoid underlining these beliefs which arise from restricted and distorted understandings of what the enterprise of science actually is. Young children do not have enough experience of scientific activity and its consequences to form opinions and attitudes towards science. If they seem to hold such attitudes it is a result of accepting adult prejudices and parroting views which are not their own.

At the primary level the concern is to give children experience of scientific activity as a basis for a thorough understanding, which will only come much later, of what science is and is not and of the responsibility we all share for applying it humanely. It is not the time for addressing the issues of rights and wrongs in the use of science. Therefore attention is given here only to those attitudes which we might call the attitudes *of* science, those which support scientific activity.

Many attitudes fall into this category and would also fall into the category of supporting learning in several subject areas. Perseverance is one of these; it is certainly needed in practising science but it is equally relevant to learning a foreign language or writing a poem. The generalised nature of attitudes is such that no clear line can be drawn between 'scientific' and other attitudes, but the ones chosen for discussion here are particularly relevant to developing ideas through exploration of the world around. They are: curiosity, respect for evidence, willingness to change ideas and critical reflection.

The importance of scientific attitudes.

It is interesting that attitudes do not feature in the national curriculum for science, a decision which can be traced back to the recommendation made in the Report of the Task Group on Assessment and Testing (DES, 1988c) that 'the assessment of attitudes should not form a prescribed part of the national assessment system.' (paragraph 30). Although this was not meant to exclude attitudes from the curricula, it did indeed result in just this. However the Non-statutory Guidelines redress the balance to some extent:

> Pupils' attitudes affect the willingness of individuals to take part in certain activities, and the way they respond to persons, objects or situations. Willing participation is an important ingredient of effective learning. The following attitudes and personal qualities are important at all stages of science education:
>
> * curiosity
> * respect for evidence
> * willingness to tolerate uncertainty
> * critical reflection
> * perseverance
> * creativity and inventiveness

* open mindedness
* sensitivity to the living and non-living environment
* co-operation with others.

These attitudes develop through encouragement and example. (NCC, 1989 p. A8)

Development of scientific attitudes

We confine the discussion here to those four attitudes which we argued above are most closely concerned in the matter of taking note of evidence and accommodating existing ideas to it.

The first, *curiosity*, is perhaps one which could be considered in the 'general' category. However, its importance for young children's learning is such that it deserves inclusion here. Curiosity leads a child forward into new experiences and so is essential for learning from exploration of things around. It follows that for children to benefit from opportunities provided for first hand investigation, their curiosity should be encouraged. To do this we need to know about how it develops and how to encourage it.

Children apparently vary enormously in their curiosity; some always wanting to ask questions and seeming never to be satisfied, following one question with another; others seem 'switched off' and don't show interest in anything. We may, however, be making hasty judgements if we use only the evidence of the questions which children ask. Some children have been deterred from asking questions by previous experience (of being told not to do so or just by never achieving satisfaction from a response, for example), others may be unsure of what is appropriate behaviour, whilst others again may be just shy and reserved. Such children need positive encouragement which legitimates expressions of curiosity, not just in terms of asking questions but through other ways of getting to know; such as touching, watching intently, using books.

The curiosity of young children, once released, is inevitably immature. It is spontaneous and impulsive, easily stimulated by new things but just as easily distracted by something else. A child who has reached a more mature level shows greater powers of concentration and will be less impulsive. The number of questions asked is likely to decline but the ones which are asked will be more perceptive and relevant. There will be more thought behind them in the attempt to equate new experience with previous knowledge. Mature curiosity shows in wanting to 'come to terms' with each new experience and reach an understanding of it. Thus curiosity becomes an active component in learning with understanding.

It is useful to sum up in terms of some general indicators of children showing curiosity:

- noticing and being attracted to new things
- showing interest through careful observation of details
- asking questions of all kinds including those which seek explanations
- spontaneously using information sources to find out about new or unusual things.

Respect for evidence is central to scientific activity. Although many new ideas have been born as a leap of the imagination, they would have a short life unless they can be shown to fit evidence and help make sense of what is already known to happen. Children's keen desire for things to 'be fair' provides a basis for respecting evidence. They will readily challenge each other to 'show me' and not be prepared to accept something as true unless they see evidence for themselves. There are many reasons why this behaviour does not carry over to their relationships with adults, and particularly teachers. Apart from status, one of the worrying reasons is that adults themselves appear to accept statements without questioning the evidence and expect children to do the same. So children build up an attitude of acceptance that what 'authority' says is true. Of course, it is impossible for evidence to be obtained for every statement and it is as immature to accept nothing as it is to accept everything. The sign of mature respect for evidence is willingness to place one's own ideas under test in relation to evidence, in the understanding that any ideas which are worthwhile stand up to such testing. As a corollary the understanding develops that no ideas are worthwhile unless the evidence is there to support them.

Actions which, if they form a general pattern of behaviour, indicate respect for evidence include:

- reporting what actually happens even if this is in conflict with expectations
- querying and checking parts of the evidence which do not fit into the pattern of other findings
- querying a conclusion or interpretation for which there is insufficient evidence

- treating ideas or conclusions as provisional and as being open to challenge by further evidence.

Implicit in the use of evidence is the *willingness to change ideas* if they are not consistent with the evidence. This attitude is sometimes described as *flexibility* but it is not to be mistaken for adopting whatever is the current way of thinking and having no ideas of one's own. At all times ideas are changing with new experience. For example, recently identified holes in the ozone layer have extended many people's understanding of ecology and indeed introduced a new concept, of 'ecocide'. However, for young children the rate of experience of new phenomena and events is particularly high and there need to be frequent adjustments in their ideas.

Unless there is a willingness to change ideas then there would be devastating confusion as new experiences conflict with existing ideas. The importance of legitimating these changes, making them open and acceptable and considered as normal, cannot be overemphasised. It is helpful to discuss with children how their ideas have changed, and to give some examples of how other people's ideas, including those of scientists, have changed. With maturity, and bolstered by greater experience, ideas need to change less often but it is important to retain the possibility of change and the tentative nature of any ideas. This can be facilitated by expressing conclusions in terms of the evidence available. 'So far we have found that all the pieces of wood float', lays a better basis for accommodating evidence that some wood (lignum vitae and ebony) does not float.

In summary, indications of children's willingness to change ideas include:

- being prepared to change an existing idea when there is convincing evidence against it
- considering alternative ideas to their own
- spontaneously seeking alternative ideas rather than accepting the first one which fits the evidence
- realising that it is necessary to change or give up existing ideas when different ones make better sense of the evidence.

In the context of science activities *critical reflection* increases the potential learning from experiences and class activities. It manifests itself in deliberate review of the way in which activities have been carried out, what ideas have emerged and how these could be improved. It is the beginning of reflecting on one's learning, but only the beginning, for this is a mature activity, requiring a degree of abstract thinking not generally available to young children. But as everything has a beginning at an earlier level than that at which it shows in its full form, we should begin to encourage critical review as a normal part of work. This requires teacher guidance at first; making time to talk through activities, to compare different approaches and to make suggestions of how thing might, with hindsight, have been tackled more effectively. The emphasis has to be on arriving at better ways of investigating or collecting evidence which can be used in the future rather than criticising what has been done.

A more mature form of this attitude shows when children themselves take the initiative in reflection on what they have done and realise the pros and cons of various alternatives. Reflecting on processes of thinking does not come readily to young children, but a useful start can be made at the level of specific activities rather than through discussion of general approaches.

So, the kinds of actions of children which indicate this attitude include:

- willingness to review what they have done in order to consider how it might have been improved
- considering alternative procedures to those used
- identifying the points in favour and against the way in which an investigation was carried out or its results interpreted
- using critical reflection of a previous investigation in planning and carrying out a later one.

CHAPTER 6

Discovery, enquiry, interactive, constructive learning – what's the difference?

Education in all areas of the curriculum, and science is no exception, is continually changing in response to development of ideas about the philosophy of the subject, the psychology of learning and the aims of education. Various terms are used in science education to describe the approaches which have been developed and held sway for a period of time. The approach which is favoured in this book has so far been described as child-centred and it might equally be called learner-centred. It is based on a constructivist view of learning but, as will be discussed later in this chapter, not all applications of constructivist learning are the same. The key factor is the extent to which the learner's own ideas are genuinely used.

History shows, however, that, in as complex a matter as education, there is no single solution. Often a combination of approaches brings the most effective education. Slavish adherence to one approach, however useful it seems in some cases, generally spells its downfall. The objectives of learning are various and so should be the approaches to teaching. In order to select and combine them it is important to distinguish between the different kinds of learning which have been advocated in the last few decades and particularly to look at their strengths and weaknesses.

Discovery learning

Discovery learning emerged from the desire to give pupils the excitement of finding things out for themselves, following in the footsteps of scientists. It was a reaction to the formal transmission of knowledge from teacher to pupil in which there was no room for the experience of participating in science. Discovery learning is also known as the heuristic method and was advocated as long ago as 1889 and 1890 in reports produced by a committee which included

Prof. H. E. Armstrong. It was a considerable innovation to suggest that the results of an experiment were to be derived from reasoning and evidence:

> The reports directed attention to the overwhelming importance of methods of instruction by insisting that mental training and the formation of accurate habits of observation, of work, of reasoning, and of description were at the early stage of education of far greater moment than the accumulation of facts or the ability to answer examination questions on these facts. The deliberate intention to achieve these ideals must lie behind all teaching of elementary science.
>
> (Report of a British Association Committee, published in The School World, September 1908, p. 424)

The aim was for pupils to be involved in the investigation of a relationship, such as that describing the balance of forces when an object floats, to gather data for themselves and to use it to 'discover' the principle which applies. This heuristic approach was reflected in the Nuffield science projects developed for secondary schools in the 1960s. It has been criticised in the following terms:

> Secondary school pupils are quick to recognise the rules of the game when they ask 'Is this what was supposed to happen?' or 'Have I got the right answer?' The intellectual dishonesty of the approach derives from expecting two outcomes from pupils' laboratory activities which are possibly incompatible. On the one hand pupils are expected to explore a phenomenon for themselves, collect data and make inferences based on it; on the other hand this process is intended to lead to the currently accepted scientific law or principle.
>
> (Driver, 1985, p. 3)

Whilst it is easy to see the value in the approach, since engaging in the methods of science is an excellent way of learning them, the force of Driver's criticism is clear when the approach is used inappropriately and as the sole teaching method. To meet some of the objections, particulary that which suggests that children are unlikely to make the 'right' discoveries for themselves, the notion of 'guided discovery' was introduced. Unfortunately pupils see through this device easily and soon react in the way which Driver suggests, thus nullifying much of the intended benefit of the approach.

Discovery learning makes no explicit reference to pupils using their own ideas. Indeed it was built on an implicit assumption that careful observation and seeking for patterns in findings (that is, the use of induction) would lead to the 'discovery' of laws and principles. This view no longer finds support either philosophically or from practical evidence. It is recognised that preconceptions determine what anyone takes notice of; it is impossible to gather all data completely openmindedly, and without some purpose there is no basis for selection. On the practical side, the problem is that children's manipulations and measurements are too rough in many cases for exact relationships to be found from experimental work. Thus they are thrown back on being told what should have happened or otherwise left with contradictory ideas.

It is a pity that 'discovery' is a term used loosely to describe all non-traditional approaches to science teaching, since the criticisms which apply to it cannot fairly be directed at other approaches which were in some cases devised to meet the deficiencies of 'discovery'.

Enquiry learning

Enquiry learning can also be called 'problem solving' since it starts from a problem which pupils are expected to solve through practical work or observation. As with discovery learning, there is a modified version, known as 'structured enquiry' in which the pupils receive suggestions for the procedures to use, how to gather data, organise it, and a series of questions which 'lead' to the solution of the problem. In the unstructured version only the problem is posed and the pupils devise procedures for themselves.

There are two main difficulties with this approach. The first is that many of the problems lead not so much to scientific activity as to what is now seen to be technological activity. This is probably most clear at the primary level. Making a bridge from paper, a candle clock which triggers an alarm, a sign which can be seen at night – will all involve the application of scientific ideas but perhaps not the development of these ideas if the activity stops at the point of solving 'the problem'. The second point is about the origin of the problem. The whole device of having a problem to solve is that it is motivating and engages intellectual activity. But if the problem is not of interest to the pupil, or may not even be seen as a problem at all, the value of the device is lost.

At the primary level, much emphasis is placed upon children working on their own problems in order to avoid the work seeming artificial and unimportant. This is much more in line with the view of science as being about problem-posing rather than about problem-solving. Thus the emphasis moves from finding problems for children to solve to finding ways of engaging children in interaction with things around them so that they find their own questions to try to answer.

Interactive teaching

This style of teaching, sometimes known as a 'children's questions' approach, was devised so that children would ask and then investigate their own questions (Faire and Cosgrove, 1988). Although children ask questions in any reasonably open activity, these questions will be so diverse and often vague that they are unlikely to become the focus of group or class activity unless the teacher takes special steps to gather, sort and convert them into specific activities. The interactive teaching approach specifies these steps and provides a structure for a science lesson which involves the gathering and consideration of children's questions as a central feature. The lesson sequence is given in Figure 6.1.

The 'Before views' step is when the teacher asks children what they already know or think about the topic. These ideas can later be compared with the 'After views' collected at the end of the investigations, but they also provide sources for investigation in their own right. The 'Exploratory activities' are intended to provoke thinking and stimulate children to ask questions. These are not formal activities in which children are expected to make written records but rather occasions for discussion between pupils and between teacher and pupils. During this time the teacher is gathering questions which perhaps are listed on the board. This leads into the 'Children's questions' step, when the children are asked to add to the questions, perhaps during class discussion, but perhaps through devices such as a 'questions box'. The questions generally need clarification, with children being asked to think about how to say more clearly what they want to know.

When the investigable questions (see Chapter 2, p. 14) have been gathered together, the next step is to select those which are to be investigated. The selection might be made by the class as a whole, so that each group works on all the chosen questions at some point, or groups may choose their own questions to work on. The teacher helps the children plan how to investigate their question, to collect resources, make their observations and prepare a chart of their results for display. The children's 'After views' are collected and compared with their initial ideas. The final step of 'Reflection' is an

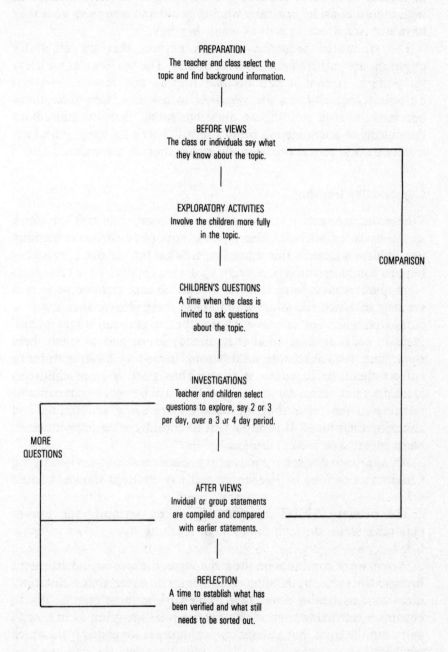

PREPARATION
The teacher and class select the
topic and find background information.

BEFORE VIEWS
The class or individuals say what
they know about the topic.

EXPLORATORY ACTIVITIES
Involve the children more fully
in the topic.

COMPARISON

CHILDREN'S QUESTIONS
A time when the class is
invited to ask questions
about the topic.

INVESTIGATIONS
Teacher and children select
questions to explore, say 2 or 3
per day, over a 3 or 4 day period.

MORE
QUESTIONS

AFTER VIEWS
Invidual or group statements
are compiled and compared
with earlier statements.

REFLECTION
A time to establish what has
been verified and what still
needs to be sorted out.

Figure 6.1 (Faire and Cosgrove, 1988, p. 16)

important one, when children are encouraged to share their ideas with others, consider critically what they did and recognise what they have not found out as well as what they have.

The structured sequence of steps ensures that the children's questions are gathered and taken seriously. The problem, as with any set pattern, is that it can become routine and lose its essential purpose. Once children are *required* to ask questions their focus becomes the need to think up questions rather than to think about the objects or phenomenon being studied. This is an approach which it may be best to use occasionally rather than as a routine.

Constructive learning

Where the 'interactive teaching' approach used children's questions as the basis for activities, the various types of constructive learning use children's ideas in this central role. What the various approaches have in common is that they begin by asking children to express their ideas about what is being studied. Where the approaches diverge is in the way in which the information about these ideas is then used.

It is possible, for example, to use the information about pupils' ideas in order to note what they already know and so teach them something new, or to note what 'wrong' ideas they have, in order to correct them, or to confront them. Thus starting from children's ideas does not necessarily mean that they will be used in constructive learning in the sense that the children are using and testing and changing their ideas. It is important to consider what happens next, when ideas have been expressed.

An approach devised for secondary science teaching devised by the Children's Learning in Science Project has the steps shown in figure 6.2.

The primary SPACE project has taken seriously the use of children's ideas, through the cycle of teaching steps shown in figure 6.3.

Where work continues on the same topic, the second and fifth steps become the same, both being concerned with ascertaining children's ideas at a particular time. This framework is loose enough not to become a restricting routine. No precise rules are given as to how to carry out the steps, but a range of techniques is suggested from which teachers choose according to the subject matter, the children and other contextual features.

The reason for 'providing opportunity for exploration and involve-

Orientation: eliciting ideas
Restructuring of ideas: involving: clarification and exchange, exposure to conflict situations, construction and evaluation of new ideas.
Application of ideas
Review of change in ideas

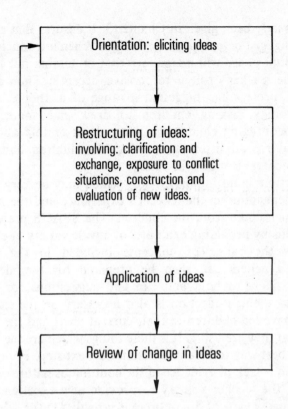

Figure 6.2 (Children's Learning in Science Project Information Sheet, 1987)

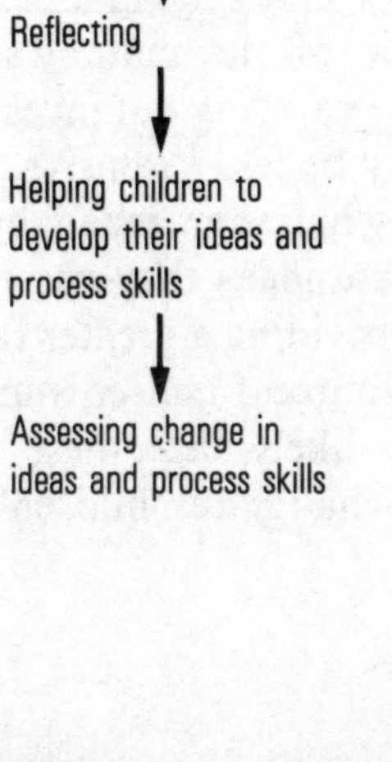

Providing opportunity for exploration and involvement
Finding out ideas
Reflecting
Helping children to develop their ideas and process skills
Assessing change in ideas and process skills

Figure 6.3

ment' has already been given in Chapter 3. It ensures that children's thinking is engaged before ideas are sought. When introducing a new topic this initial phase will merge into that of 'finding out ideas'. In this phase the teacher's role is responsive, accepting, non-directive. The SPACE project has suggested a range of methods, including open questioning, asking children to draw and annotate their drawings, encouraging children to write about or talk about their ideas, and above all listening to what the children have to say (SPACE Teachers' Handbook, 1992).

The 'reflection' is an important step which may be very short or quite long, depending on the teacher's experience and the nature of the ideas which come from the children. The SPACE project helps teachers in this by providing examples of a wide variety of children's ideas across the range of concepts included in the national curriculum. Teachers can then be prepared for the ideas their children are likely to hold and can find suggestions for ways of responding to them. Reflection is also necessary on the experience which may have led children to their current ideas and the kinds of reasoning that they are using. It is these things which are the basis for deciding the best way to help children in the next step. For example, if the problem is lack of evidence of the continuous nature of growth of seedlings, the teacher will try to provide new experience which would challenge a view of discontinuous growth. On the other hand, if a child's ideas are highly bound to particular contexts (such as having different explanations for evaporation in different situations) it may not be new evidence which is required, but a discussion of the reasoning, or perhaps the words, being used about evidence already available.

For 'helping children to develop their ideas' the starting point is the ideas which the children have, whether or not these ideas are along the 'right' lines. 'Development' means extending and strengthening useful ideas as well as challenging ones which are not useful in explaining things. One of the main ways of doing this involves children in devising and carrying out investigations to test their ideas, but some activities may be less extensive, such as asking children to provide examples of what they mean by certain words they use, encouraging them to use ideas they use in one context in trying to explain another one, providing a greater range of evidence (including the use of secondary sources) and encouraging reflection and communication. It is very likely that these activities will provide the opportunities for any change in children's ideas to become evident and thus the next step, of 'assessing change in ideas' is automatically covered. If not, some of the same techniques as used in the 'finding out' step need to be repeated.

CHAPTER 7

National guidance to learning experiences

Prior to the introduction of the national curriculum in England and Wales, teachers and schools had almost unlimited freedom to decide what was taught at the primary level. In theory the same freedom was there for secondary teachers, but in practice the examination syllabuses largely dictated pupils' learning experiences. The statutory control of the curriculum was placed in the hands of the local authorities under the 1944 Education Act, but this power was not used and control devolved to the schools. Local Authority advisers and inspectors often issued guidelines and provided courses which steered parts of the curriculum in certain directions, but these were never presented as regulations to be followed.

The 1988 Education Act (DES, 1988a) returned to the Secretary of State powers to prescribe the curriculum in England and Wales similar to those which were provided under the 1902 Act and which were changed only by the 1944 Act. As a result, in England and Wales, the *National Curriculum* has been created and it is required by law that it is taught. In Scotland the new curriculum is called the *5 – 14 Development Programme* and is non-statutory, but continues a long tradition of central curriculum guidance which contrasts with the decentralised curriculum in England and Wales.

This chapter is not an exposition of the national curriculum and its equivalents in Scotland and Northern Ireland, since this is best obtained from the official documents. Instead its purpose is to set the new developments in context and to point up some issues affecting the teaching of science.

What is and is not prescribed by the national curriculum

The existence of a national curriculum is indeed a great change. The greatest and most significant for primary school is probably the requirement for science and technology to be taught from the age of 5 and for science to have the status of a 'core' subject.

But the extent of change in terms of curriculum content has to be considered in relation to what is dictated and what it remains for teachers and others to decide in relation to learning experiences. The kind of curriculum statements contained in the national documents in the UK are confined to stating the objectives of learning (the attainment targets) and what knowledge, understandings and skills should be taught (the programmes of study). It is explicitly stated that the Secretary of State may not prescribe the time to be spent on any programme of study, how provision should be made within a school timetable for subjects to be studied nor what teaching methods or materials should be used (Circular 5/89). Thus a primary school can decide to adopt an integrated day or a subject based timetable, as it prefers, it can embed experiences in free activity or structured exercises, it can decide whether or not to base work within topics and if so, which topics – providing the pupils encounter the experiences indicated and work towards the attainments set out. There is guidance about methods and approaches, however, in the Non-Statutory Guidance provided by the National Curriculum Council (NCC) and its equivalent in Wales. These documents, written by groups of teachers and advisers, contain firm and useful advice but, regrettably, in a form which is not very accessible to teachers.

Thus there is strong guidance for teachers who wish to follow it and indeed the particular statements of attainment do limit teachers' freedom to teach in any way they like. It is hardly possible, for example to work towards achievement described as 'be able to group materials according to their physical properties' without some first hand experience of the properties of materials. Indeed a major purpose of the introduction of the curriculum was to produce a greater degree of uniformity in teaching with the inevitable consequence of reducing the freedom of individual teachers. Looked at positively this was in the interests of the pupils so that there would be greater compatibility in the curriculum of one school and another. Pupils moving from one primary school to another would not experience such great changes as had been the case when each school decided what to teach without any overall common framework. Furthermore, the introduction of a national framework required teachers to develop whole school plans for implementing it, thus improving continuity from class to class within a school.

Approaches to curriculum control at the national level

Even within the UK there are considerable differences in the extent of what is statutory and in the nature of guidance. Thus it is worthwhile taking a brief look at how different countries interpret and implement a national curriculum.

Some of the main ways in which curriculum control may be exerted at the national level are by instituting

— a national curriculum which is statutory
— a national curriculum which is advisory
— national testing for all pupils at some ages
— national testing which samples pupils at some ages
— approved text books and books for teachers
— inspections of school.

Practices vary widely as a few examples will show. Japan, for instance, has a statutory national curriculum, national testing and approved textbooks which are used in every school. There are tests for entry to secondary school which exert a further strong influence on the primary curriculum. In Germany, compulsory curricula are prescribed at the regional level but, as in the UK, these stop short of prescribing methods and materials. There are no national or regional tests in primary schools or at transition to secondary school. In the United States control is exercised at the State level and practices vary widely. In many States testing is the dominant tool of control rather than curriculum statements and text books. Testing tends to override other forms of control in most situations and where it is frequent, what is tested dictates what is taught rather than vice versa.

In none of the countries of the UK are there nationally approved text books, which would be inconsistent with the rejection of prescribing methods and materials. The curriculum control is through specifying objectives to be achieved and requiring that achievement be assessed and reported regularly. The particular form of these requirements varies among the countries of the UK, however.

In Scotland, the 5 – 14 Development Programme, which is not compulsory, extends, as its title implies, only to the age of fourteen. The curriculum is divided into five broad areas: English language, mathematics, environmental studies, expressive arts and religious and moral education, (SOED, 1990). For each of these, statements identify five levels of attainment for each of a series of 'strands' within major headings of attainment outcome. These five levels are defined in terms of what most pupils at various levels in the primary school should be able to attain. The Scottish 5 – 14 Programme does not set out programmes of study but uses this term for the plan which the school draws up for the curriculum which its children will study. Some 'key principles' are provided for schools' guidance in this process, relating to balance across the components of the curriculum, to matching teaching approaches and contexts to what is taught, to catering for differences of attainment within a class and to providing equal opportunities to learn for all pupils.

Whilst the curriculum statements are described as 'guidelines', teachers are required by law to carry out regular assessment at all ages and testing at the ages of 8/9 and 11/12 (P4 and P7) in language and mathematics only. An optional report card has been produced for reporting children's performance to their parents. It is clear, therefore, that for three of the five areas of the curriculum the level of attainment reported will depend entirely on the teacher's assessment.

In Northern Ireland, the curriculum is expressed in the same form as the national curriculum for England and Wales, adhering to the same ten levels of attainment and defining the curriculum in much the same subject-based terms. Arrangements for testing are similar to those in England and Wales but in addition there are plans for externally developed test materials 'to support and inform the professional judgement of teachers' (NISEAC, 1990, para 22).'

The matter of assessment for various purposes and the relationship of assessment to the curriculum will be taken up in Chapters 20 and 21. For the rest of this chapter we return to the guidance which the national curriculum in science for England and Wales provides a basis for learning experiences at the primary level which advance children's ideas through the use of process skills.

Children's learning in science and the national curriculum

It may be helpful to list some of the significant features of the science curriculum in its written form which affect its translation into practice:

(1) It is subject-based, so science is identified separately from other subjects.

(2) In the attainment targets, processes are separated from aspects of knowledge and understanding.

(3) For all attainment targets progression in achievement is described in terms of ten levels of attainment.

(4) Programmes of study are identified for each key stage and broadly outline the experiences children should have in order to achieve the objectives expressed in the statements of attainment.

The first of these features results from using the same structure for the curriculum throughout the primary and secondary years up to the age of 16. The continuity in structure helps in ensuring a coherent development within a subject but it is more in keeping with the usual arrangement in secondary schools than in primary schools. Primary schools therefore have to take extra steps in implementing the curriculum if science is to be integrated with or even related to other subject experiences. There is a clear danger of the simplest path being to teach science and other subjects separately from one another as this makes the curriculum easier to cover. It is noteworthy that the Scottish 5 to 14 Programme has avoided this by presenting the curriculum in broad areas, with science as part of environmental studies (in which history, geography, technology and social studies will have a part). Some of the ways in which primary schools have tackled the planning of science within a topic based or integrated subject approach are discussed in Chapter 19.

The separation of processes from the knowledge and understanding attainment targets is another feature which means that the curriculum requires considerable thought to become reshaped in practice. The non-statutory guidance makes clear that there is to be more or less equal emphasis on process and knowledge and understanding. But the fact that there is one process attainment target against several for the knowledge and understanding component militates against this equality. Despite all the good advice in the non-statutory guidance, it is not unknown for the processes to be treated in the same way as content and 'taught' at a particular time or in one topic rather than being used in all activities in the way required for the learning described in Chapter 2.

Part of the problem of planning effectively for the use and development of process skills stems from their description in the ten levels, which brings us to the next point. In relation to the knowledge and understanding attainment targets, the mapping of different levels of sophistication in the understanding of a concept was a marked step forward, even if the steps identified may not be all that accurate. It provided an explicit statement of how primary work forms a foundation for ideas which are to be further developed in the secondary years. It also suggests the limited objectives for primary science. Primary teachers at last have an explicit statement of the levels of basic understanding to which they are expected to work and many have been relieved to find that these are not the abstract ideas which they half remember from their own secondary school science.

For the process skills the statements give an impression that the progression outlined in the ten levels is independent of the subject matter in which the skills are used. Any skill has to be used in some context and that context affects the use of the skill. Young children may, for example, be well able to plan a fair test in relation to which of two paper aeroplanes flies the furthest, but not in relation to which of two solutions has the greater osmotic pressure. Obviously the familiarity of the subject matter and the complexity of the concepts determines the extent to which skills can be deployed successfully. With less extreme differences in demand of the topic there is still a considerable problem in deciding at just what level a child is operating in relation to process skill development.

The very broad terms in which the programmes of study are expressed leave the teacher the freedom to choose particular subject matter which fits the interests and locality of particular groups of children. For example the requirement that 'Children should work with a number of different everyday materials grouping them according to their characteristics, similarities and differences' leaves it to the teacher to decide the particular materials to be studied as well as when and how this should be done. Indeed the lack of structure of the Programmes of Study, which do not prescribe even in general terms what is to be studied in each year but only for the four years of the junior school in the case of Key Stage 2, may be the reason why many teachers use statements of attainment in their planning instead of the Programmes of Study.

CHAPTER 8

Evaluating, selecting and adapting learning experiences

As was made clear in Chapter 7, the statutory national curriculum says nothing about teaching methods and materials. Neither are there recommended books for teachers or pupils. The teacher has, however, to decide these things, to choose between the many different versions of classroom activities. On what basis should this choice be made? What criteria should be used in deciding that one approach is more worthwhile or appropriate than another? These are the questions which are taken up in this Chapter. The concern is not with the subject matter, which it can be assumed will be consistent with the development of knowledge and understanding in the national curriculum, but with the way in which children are brought into interaction with it.

Knowledge and understanding cannot be passed to children directly. All that was said in Chapter 2 indicates that understanding is not developed by rote learning: for example, children do not come 'to know that light passes through some materials and not others, and that when it does not, shadows may be formed' simply by being told this and asked to memorise it. So much is agreed. But in providing more genuine opportunities for learning there are still many aspects to be decided, for instance:

• The question of whether the science should be studied in the context of real life events which provide interest and stimulate curiosity but are complex and present ideas often intertwined together, or in 'tidied up' activities which demonstrate scientific relationships and concepts in a more direct form but which then have to be related to real situations.
• The role of the learners in addressing their own questions or those proposed by someone else.
• The extent to which various of the process skills should be used

PLAYING WITH SHADOWS
Prop a piece of white card upright on your
desk.
Shine your torch on the card.
Stand a pencil upright between the torch
and the card.
What do you see on the card?
Hold the torch still and move the pencil.
Which way does the shadow move?
Stand the pencil four inches in front of the
card.
How high is the shadow?
Move the card back until it is eight inches
behind the pencil.
How high is the shadow now?
Is it larger or smaller?

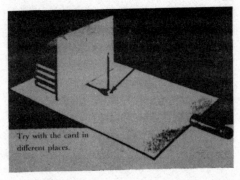

A SHADOW THEATRE
Cut the shape of a man out of card.
Hold it behind the sheet.
Can you make the shadow move?
Write a play for your shadows to act.

(From *Working with Light*, Catherall and Holt, pp. 18, 19)

Figure 8.1

and developed in the experiences. It is accepted that process skills
must be developed as much as knowledge and understanding, but
how this affects decisions about each activity is not clear.

Many of these points are part of a wider question of the extent to
which an activity contributes to an understanding of scientific
activity and to the development of scientific attitudes.

Criteria in action

It is useful to have some examples in mind in order to reveal the
implications of different sorts of activities. The three examples of
activities in figures 8.1, 8.2 and 8.3 are all about shadows and could
help to develop ideas about the reason for shadows and to demonstrate that light travels in straight lines.

You will need

A darkened room or cupboard
A good torch

What to do

Shine a torch on the wall.
Hold a finger in the beam of light.
What do you see?

The finger blocks part of the light and creates a shadow.
Use your hands and fingers to make shadow animals.

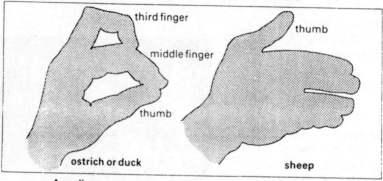

An eclipse

Sometimes, the Earth moves between the sun and the
moon.
The Earth's shadow falls on the moon.
We call this an eclipse of the moon.

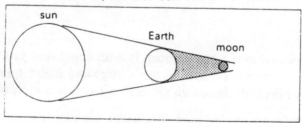

Figure 8.2 (*Science Horizons*, Shadows, Activity 11)

It is useful to think about which of these approaches would be your
first choice, if you had to make a decision, and which would be your
second choice. The reasons for the choice are more significant.
Perhaps none of the examples would be favoured as a whole because
there are things that are disliked about each one. There are presuma-
bly things that are liked about each one also. Taking each one and

---

MAKING SHADOWS

On a sunny day take a variety of objects and put them in sunshine and use them to cast shadows.

What different shadows can be made with the same object?
What different shadows can be made with different objects?
Does the time of day make any difference to the shadow?

Make a shadow on a wall with a hoop.
   How many different shapes can you make from it?

Make a shadow from a round plastic litter bin lid.
   How many different shapes can you make from it?
   When are the shadows of the hoop and lid the same and when are they different?

Put various articles in front of a torch, but before you switch on take turns in the group to say what you think the shape of the shadow will be. Why do you think it will be this shape?
If you are not right, try to work out why and improve your predictions.

Make up a game which is played by moving shadows.

Figure 8.3

deciding what is liked and disliked about it is another way of identifying the criteria used in selection.

The following criteria for evaluating activities were suggested by the working group which developed the national curriculum (and were later brought together in a slightly modified form in the Non-Statutory Guidelines to the national curriculum):

Will the experience
- stimulate curiosity;
- give opportunity for developing scientific skills;
- give opportunity for developing attitudes relating to scientific activity including cooperative working;
- give opportunity for developing basic scientific concepts;
- relate to the interests of children at a particular age and to their everyday experience;
- appeal to both boys and girls and to those of all cultural backgrounds;
- help children to understand the world around them through their own mental and physical interaction with it;
- involve the use of simple and safe equipment and materials;

- involve resources and strategies accessible to teachers;
- give opportunity to apply scientific ideas and skills to real life problems, including those which require a technological solution;
- offer opportunities to work cooperatively and to communicate scientific ideas to others;
- contribute to a broad and balanced science curriculum, bearing in mind other experiences already selected?

(DES, 1988b, p. 95)

These criteria can be compared with those used in considering the examples on shadows earlier. It is useful to consider those examples again in relation to these criteria and to see if the order in which they would be placed coincides with the original preference. Such an exercise brings into sharp focus the unavoidable role of values in curriculum decisions. We prefer the type of activities which reflect our values and indeed we would not teach effectively through curriculum materials which were not consistent with these values. Making them explicit is an important step in recognising and justifying the choices we make in providing certain types of learning experiences for our pupils.

Adapting activities

A further advantage of making explicit the criteria on which we evaluate and select activities is that it provides a basis for adapting and improving activities. Take the example in Figure 8.4.

There are some obvious reasons why this is limited as a learning experience although it is certainly an activity most children would enjoy. It is perhaps useful to recognise what is valuable about the activity before criticising it.

It is capable of relating to children's interests across a broad spectrum, with no obvious gender or cultural bias. It uses simple and safe materials, which are familiar and cheap and it would be an easy activity for teachers to manage.

On the other hand there are many ways in which the activity could be changed to meet other criteria. For example:

stimulating curiosity
The activity might begin with the experience of throwing several parachutes, of different sizes and even shapes, and noticing how they fall. The question as to why the differences can be raised.

Parachute

- Cut a 14-inch square from sturdy plastic
- Cut 4 pieces of string 14 inches long
- Securely tape or tie a string to each corner of the plastic
- Tie the free ends of the 4 strings together in a knot. Be sure the strings are all the same length
- Tie a single string about 6 inches long to the knot
- Add a weight, such as a washer to the free end of the string
- Pull the parachute up in the centre. Squeeze the plastic to make it as flat as possible
- Fold the parachute twice
- Wrap the string loosely around the plastic
- Throw the parachute up into the air

Results. The parachute opens and slowly carries the weight to the ground.

Why? The weight falls first, unwinding the string because the parachute being larger, is held back by the air. The air fills the plastic slowing down the rate of descent if the weight falls too quickly a smaller object needs to be used.

Figure 8.4

skill development
Opportunities for children to develop their process skills are limited by the lack of any investigation once the parachute is constructed. There are many variables which affect the fall of the parachute, such as shape, area, length of strings, which children should explore in a controlled way as they test out various ideas about why there are differences between one and another.

developing attitudes, working cooperatively and communicating
There could be instructions for pooling ideas within a group, planning how to find out 'what happens if . . .' and preparing a group report to others. At intervals in the work the children should meet together as a class to listen to reports of each others' progress and share ideas.

basic concept development
A main point of the activity is to enable children to recognise the role
of air in slowing down the fall of the parachute. With this in mind it
would, therefore, be useful for children to observe how quickly the
parachute falls when it is not allowed to open. Exploration of larger
and smaller parachutes might further children's ideas about the effect
of the air. The question of why the parachute falls at all could also be
discussed leading to a recognition of the main forces acting on the
parachute when it is falling.
Giving the 'answer' to why the parachute moves slowly is not
allowing the children to use and explore their own ideas; so this part
should be omitted.

applying scientific ideas and skills to real life
The uses of air resistance are many and not restricted to parachute
descents from aircraft. Children should be encouraged to think about
air resistance in relation to horizontal movement, in yachts and
sailing ships as well as in slowing aircraft in landing on short runways
and aircraft carriers. They can be challenged to think about the kind
of materials and construction which is needed in each case. They will
also be able to relate to more everyday events, such as riding a bicycle
in a strong wind and the 'helicopter' wings of sycamore seed seen
drifting gently down to the ground.

There are two main consequences of modifying the activity in these
kinds of ways:

• First the activity will depend more on the teacher than on a work
  card, although careful wording of worksheets or workcards can go
  a long way in encouraging children to use their own ideas and
  think things out for themselves (see Chapter 17, p. 00). The
  teacher's role in developing children's ideas, skills and atitudes is
  the subject of the next three chapters.
• Second, it will undoubtedly take up more time. This has to be
  balanced by the much greater learning which takes place. Further,
  had the same time as required for the modified activity been used
  for several activities of the original kind there would still be no
  opportunity for some of the learning which is required.

Fewer activities, with more opportunity for different kinds of
learning, for discussion and for developing skills will be a greater
contribution to learning with understanding. The use of criteria, such
as the ones quoted above will help in this judgement.

CHAPTER 9

Matching learning experiences to children

The model of learning in Chapter 2 identifies a learning experience as something with which the learner can engage using existing ideas and process skills. Thus some form of matching is assumed between the point that has been reached in relevant ideas and skills and the demand required to understand the new events or objects. In the classroom context, where the learning experiences are those provided through the curriculum, this means two things: a careful choice of content and adjustment of the kind of interaction with it that is expected.

These two aspects have to be considered together because there is generally the opportunity to engage with a certain content at a variety of levels. Take for example, floating and sinking. Figure 9.1 shows just a few of the many questions which can be pursued with the simplest of equipment.

Which of these activities is embarked upon will depend on the ideas already available to the children from previous experience; once started, each can lead on to many other things. The teacher clearly plays a very important part in selecting the level of demand of an activity, as we shall see later.

Matching can be thought of as requiring decisions at two levels. First is an overall *macroscopic* matching, where activities are provided at the right level of demand for the whole class. This means avoiding with young children, for example, experiences which could only be understood in terms of a concept such as osmotic pressure. Second is matching at a *microscopic* level, where each child is able to engage with the subject matter in a way which advances his or her own understanding. This is also known as *differentiation* and means 'fine tuning' of the demand of the activity to suit the needs of individual children. We shall consider these two levels after first making clear the meaning of 'matching' as it is being used here.

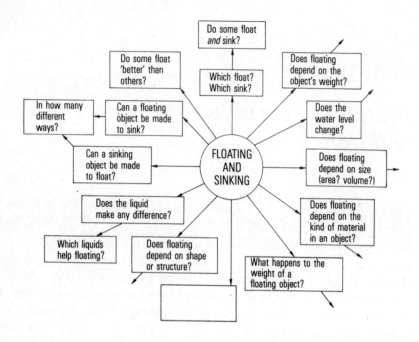

Figure 9.1 (*Match and Mismatch: Raising Questions*, 1977)

Matching

Matching is finding the balance between presenting something which is so far away from previous experience that children cannot engage with it with their existing framework of ideas; and presenting something which is too simple and familiar to test and extend ideas and skills. It is by no means a new idea and was well captured in the CACE Report Children and their Primary Schools in 1967:

> If the material is too familiar or the learning skills too easy the children will become inattentive and bored. If too great maturity is demanded of them, they fall back on half-remembered formulae and become concerned only to give the reply the teacher wants. Children can think and form concepts so long as they work at their own level and are not made to feel that they are failures.
>
> (CACE, 1967, paragraph 533)

This sums up well the damaging effect of mismatching. When this happens the problem is not only what children don't learn but what they do learn; they learn that school is a boring place or that it is a place where they are bewildered and never able to do what is expected of them. Repeated experience of failure to understand may set up a vicious circle in which both teacher and child reach a stage of never expecting anything but further failure, which then becomes inevitable. Take Duncan, for instance:

> Duncan had never been any good at maths and would often say so to various people. As seven he could get by if he was told what to do – and his neighbours would oblige by showing him what they had done. But at eight he was distressed by having to do maths work and frequently burst into tears at the sight of his maths book. Help with particular difficulties sometimes encouraged him but he was never able to tackle new problems for himself. One day another teacher was helping out and sitting at Duncan's table and he plucked up courage to ask her 'What does that mean?' 'That' was the figure 8. He knew it was called eight but he didn't seem to have connected it with eight of anything. His own teacher was then able to appreciate what was perhaps at the root of his problem with numbers and helped him to understand how sums on paper can represent actions with objects. Then he began to succeed for the first time with simple addition and subtraction.
>
> (From *Match and Mismatch: Raising Questions*, 1976)

It is indeed rather easier to be definite about the problems of mismatching than about the possibilities of matching. This is because matching is not a static condition, but a *dynamic* process in which the teacher's role is paramount. It depends on the teacher finding out and using information about children's ideas and skills and adjusting the activities to provide the opportunity for development. This may sound impossible to do for every child, but it does not require a different activity for every child nor require children to work alone. These conditions would sever important learning opportunities and be counterproductive. The distinction between macroscopic and microscopic levels of matching is useful here.

Macroscopic matching

We must remember that a child's learning environment extends beyond the bounds of the classroom and school to the home, the local neighbourhood and, through the media, to the national and international context. There are many aspect of this total learning environment which can affect learning, directly or indirectly, but are outside the influence of the school. What the school has to offer has to be considered within this wider setting.

Generally schools provide a classroom setting, equipment and activities which are matched to the children at the macroscopic level. For infants there is plenty of opportunity to

- explore materials and objects provided in the classroom, using all senses as safe
- try things out
- do things over and over again
- look at things in the natural environment around the school
- pour, build, make, sift, bake, plan
- take things apart and put them together again
- talk, write and draw about what they have done and what their idea are about it.

For lower juniors the curriculum provides opportunities which match their developing skills and ways of thinking and enable them to

- have experience of finding answers to their own questions
- explore a wider range of materials and objects both within and outside the classroom
- test out ideas through investigations where variables are manipulated so that tests are 'fair'
- observe events and sequences in detail and with the aid of instruments
- seek patterns in their findings
- find simple explanations for some everyday phenomena.

Upper juniors, with greater powers of concentration, precision in manipulation and measurement and ability to plan ahead, have opportunity to

- undertake extended investigations or problem solving which advances their ideas and skills
- plan investigations and discuss plans critically before carrying them out
- discuss different kinds of questions and identify those which can be answered scientifically

- review their work critically and identify the pros and cons of alternative ideas and procedures
- extend their knowledge using secondary sources of information.

The items in these lists suggest how, in broad terms, the subject matter should be treated. The choice of subject matter is indicated, again in broad terms, by the national curriculum (in England and Wales) or by other guidelines, in Scotland and elsewhere. The particular way in which a school proposes to implement these guidelines will be indicated in the overall school programme which lays down what each class will cover (see Chapter 19). In this way a teacher arrives at a set of activities which are designed to be 'in the right ball park' for a particular class.

Matching at the microscopic level

Matching for each individual child means finding the cutting edge of each one's learning; the point that really makes them think, the thing they are wondering about. It doesn't mean that they need individual activities; quite the reverse, since working in a group where there are differences in ideas is a productive way of developing ideas. What it does mean is that the demands of the activity and the expectations from it take account of a child's unique set of skills, ideas and attitudes. The following account of some work in an Infants class gives an example of this matching in action:

> Some work on area was prompted by the delivery of new tables for the classroom. Was there the same room on the top of the new tables as on the old? How much surface space was there on two tables fitted together compared with separated ones? These kinds of problem were tackled successfully by Simon's two working companions who began to use squares as a unit of measurement, but Simon himself was still doubtful about conservation when the shape or position of an area was changed, or when it was divided up. He continued to work beside the other two as they pursued their work outside, using large square units to measure the area of a circle marked out on the playground. They tied ropes together to a length equalling the circumference of the circle and then altered the shape to a triangle. Was the area of the triangle the same as that of the circle? The two other boys gave different answers: Simon agreed with both! There had been many occasions before where Simon had chosen to agree with his friends rather than think things out for himself and he didn't see anything peculiar about agreeing with conflicting opinions. Without separating

him from his companions his teacher made sure that Simon was encouraged to think about surface space, in covering boxes with paper, seeing how many different shapes he could make with tiles of a constant area, and so on. At the same time she made sure that sometimes he had to work alone and make his own decisions. When in a group she made sure that he was sometimes called upon to speak first when reporting.

(From *Match and Mismatch: Raising Questions*, 1977)

What comes through here is the teacher's knowledge of Simon (who is the focus on this occasion) and of other children. This knowledge extends far beyond the cognitive aspects of his development, for it is the knowledge of his personal characteristics which informs the decision about how to help Simon without making him feel insecure. The setting of individual goals just far enough for each child is further illustrated by this teacher of juniors who is describing how she encourages children to finish or continue with a piece of work:

But of course it sometimes depends on what one means by 'finish'. A piece of writing three lines long and a drawing is only the start for Timothy and I would expect him to continue with it. For Iain this would be a good product and I would praise it and not take the edge off his pleasure by saying that it could be more. The praise is usually enough to encourage more next time, but if not, then it would be time to say that he should try to do more than he did last time.

(Ibid)

The supportive atmosphere which is evident here is also important in avoiding the vicious circle of failure which arises when children disguise their lack of understanding by rote learning in their concern 'to give the reply the teacher wants'.

Matching at the level of individual children means that the class and curriculum organisation must allow for differentiation of experiences for the children. This can be provided in a variety of ways. It does not mean that children have to work individually, but that differences in their experiences arise from the help they receive from the teacher and the goals which are set at the level which is achievable. The various ways include:

(1) Having the whole class working on a common theme but on different aspects of it. For example it would be quite possible, and probably desirable, for children working on floating and sinking as a class but for groups or individuals to be working on

the different questions, such as those indicated
using their own ideas to tackle them.

(2) If the topic does not allow for differentiation by di.
giving the children the same task (not necessarily all at .
time) but with different degrees of teacher support for childi.
different abilities.

(3) Grouping the children by ideas in relation to the topic.

(4) Creating mixed groups so that the children can help each other,
but ensuring that some recording is done individually so that the
teacher can assess the advance made by each child.

There is more about class organisation and the pros and cons of these
suggestions in Chapter 17.

CHAPTER 10

The teacher's role in development of ideas

The development of ideas and understanding goes hand in hand with the development of process skills and scientific attitudes. The essential interdependence of these aspects of learning was emphasised in Chapter 2. Learning with understanding involves development of ideas through the learner's own thinking and action and in science this means that process skills are used and through use are developed to deal with new situations. Attitudes, being more generalised components of behaviour than process skills and concepts, depend upon being fostered in a wide range of experiences since there is no way of 'teaching' attitudes directly. This essential interconnectedness of ideas, or concepts, skills and attitudes is not denied by focussing on each one at a time in considering the teacher's role in this and the next two chapters. We are simply looking at different facets of a whole, just as we can look at different faces of a solid without turning it into something two-dimensional.

The teacher's role in helping children to develop ideas has several aspects, of which we shall consider the following:

- gaining access to children's ideas
- deciding the next steps which can be attempted
- taking action to help the development
- knowing when to stop.

It may seem that taking action is obviously the most important of these, but this step cannot be taken unless there is information about the current ideas of the children and some reflection on how these ideas can be built upon – if we are aiming for learning with understanding, that is. Then there is always the need to consider children's response to new experiences and to take a decision about whether to continue or whether to leave ideas as they are for the present.

All these aspects are equally important. Indeed it could be argued that taking action without the other three aspects being in operation could be counter-productive, since it would be difficult to avoid a mismatch between a child's development and the activities provided. *These are not sequential steps; they are parts of the on-going role of the teacher.*

Gaining access to children's ideas

Children have to be involved in relevant explorations or enquiries when their ideas are monitored. This may be at the beginning of a new topic, as in the example given below, or during the course of the work. There is a variety of techniques which can be used to enable children to reveal their ideas, which will be reviewed in Chapter 22, but by the nature of the demand for immediate information, they cannot be used for all children at the same time. For building a picture of children's ideas, teachers are bound to depend on picking up snippets of information from listening and watching children during their activities. These snippets are often uncertain and fragmentary and can be contradictory, as reported in this teacher's account:

> Before the children in my class began to build Viking ships, my enquiries revealed that none of them had any understanding about what determines whether or not an object would float. However, practical work led to a number of hypotheses, such as 'because it is made of plastic' and 'because it's round', which further investigation either confirmed or dispelled, until eventually I felt satisfied that nearly all of the class had gained a fairly clear understanding. The next requirement, in the interests of developing recording skills, was for the children to write a report of the work undertaken and to state their conclusions. As anticipated, some wrote competently while others needed encouragement; some conclusions were incomplete, but in most cases this was because the writer had 'run out of steam', so teacher intervention through open-ended questioning was called for. The reports indicated that, with the exception of wood, the children considered materials to be a less significant indicator than shape and stated that '. . . so if a thing is light for its size it will float, but if it is heavy for its size it will sink. The weight needs to be spread out'.
>
> At this point I did indeed believe that my initial assessment had, for the most part, been well founded; two cases, however, showed that the situation was not quite so straightforward.
>
> Gary, whose understanding had shone forth throughout the practi-

cal work, had written simply that 'Air makes things float.' Questioned the next morning, he said that 'A layer of air comes under things and holds them up'. This conclusion was certainly not consistent with his most recent experiences. 'If it is a layer of air like that', I asked, 'would you expect things to be wet or dry?' 'Yes', he replied, then, 'Oh aye, it can't be, because they were all wet'. He was unable to articulate, at that point, just what he thought did allow things to float, but after weighing out 100g of plasticine he immediately fashioned it into a broad-beamed, thin-walled shape, something between a coracle and a soufflé dish, which floated without any adjustment.

Melanie and Natalie had also shown a quick grasp during initial investigations, and, unlike Gary, had drawn valid conclusions in their report, referring to the relationship of shape, size and weight. What a surprise I had when they too had the opportunity of putting their understanding into practice: they weighed out their plasticine and cooperated well to produce with it a very elegant, small boat with thick sides, heavy mast, and chunky sail which, inevitably, sank like a stone. Undaunted, they discussed their results with each other and with me, and without quoting their written conclusions I encouraged them to take account of what they had hypothesised the previous day. It took them all day to get their plasticine to float, and that despite inspection of successful attempts, heralded by cries of jubilation from other groups. Their pleasure in their eventual success was evident and I gave a wry smile when Natalie said, 'It's the shape, isn't it? We had to get it spread out more so it wasn't all heavy on a little bit of water.'

My day epitomised the challenges we face, week by week, in the classroom, constantly adjusting backwards as well as forwards, our assessment of a child's understanding, and catering for the needs revealed to us . . .

(Pat Holmes, *Primary Science Review*, No 15, 1990 p. 2)

This account shows how, even in the same context, children's ideas can appear to change.

There is certainly plenty of evidence that there may be a difference from context to context. For instance, the explanation given by the same child of a phenomenon, such as the condensation of water on a cold surface, may well be different according to whether this is on the side of a cold drinks can taken from the 'fridge or results from warm moist breath on a window pane. When the purpose of finding out children's ideas is to help them, this contradictoriness is not a problem. It is in fact helpful to know that in one situation a child applies an idea or skill but not in another. It can give a clue to how to help by revealing the situation in which there is better performance. It is only a problem if there is a need to make a decision as to whether a child has or has not an understanding of a particular idea. So, as long as the purpose it to help the children and not just to make a decision about their achievement, these fragments of uncertain information picked up in the course of activities provide what is needed as a basis for taking action within the activity. Moreover, over a period of time the pieces combine to form a picture of each child; a picture which is constantly changing as the child develops.

Deciding the next step

To decide the action to take in the light of evidence about the ideas children have, it is necessary to keep in mind the starting point and the general direction towards more widely applicable scientific ideas. For example, the existing ideas might be

— *there is nothing in the empty space between objects around us*

and the scientific idea

— *air is all around us.*

But the gap between these two is not to be covered in one leap. The question arises : how far it is desirable to go in one step?

The answer to this question in a particular case will depend on many features relating to the child's interests, what motivates him or her, what materials are available, how much attention the teacher can give, and so on. Although every case will be different, there is a *general approach* which involves reflecting on what the child's ideas are and how they may have been formed. If the teacher has no ideas about the possible origin of the child's thinking then the first thing is to ask for the child to explain. The answer often is to the effect (stated either explicitly or implied) that it is 'obvious'. It may well seem obvious, for instance, that there is 'nothing' in the spaces around us. But this is not unhelpful as a clue to the basis of the child's idea, for it indicates a judgement based only on what is seen and perhaps a limited experience of events (such as not being able to pour water into an 'empty' bottle unless the air is let out) which would lead to querying the notion of 'nothing'.

Gaining some clue to the experiential base of children's ideas is one lead to possible action. Another is to give careful attention to the words children use. They often use words with a restricted or idiosyncratic meaning and by probing the meaning they have for the words it may be possible to identify action to take. Children might be asked to give examples of what they mean by a word or provided with further activities which would involve the use of the concept and the opportunity to discuss whether this fits with their use of the word. For example, the word 'nothing' might not necessarily mean a total absence of matter, as in a vacuum.

Exploring the origin of ideas may indicate whether or not they have been tested in relation to available evidence. This means using them predictively to suggests what would happen in certain circumstances and then seeing whether this really does happen. So the idea that there is 'nothing' around things would lead to the prediction that there would be nothing to stop things moving through the space. Trying to move with a furled umbrella in front of you should be no different from trying to move with the umbrella open. This can be readily tested. Other evidence is often present but has been ignored. So, for example, the swaying of trees and flags cannot be explained by a 'nothing' hypothesis and discussion of these things is another way of testing the children's idea.

The approach, then, is to explore the children's ideas and particularly the basis on which they have been formed. This basis will suggest ways of moving in the direction of the scientific view. But success cannot be guaranteed, for as has been said, there is no formula which can be followed. However, there are two further points to bear in mind.

First, that the children are partners in the endeavour to learn. Their reactions will give a strong indication of whether or not the action is appropriate. They are the ones who really know whether they are frustrated or confused, whether they want more time to get a grasp of an idea or want to move ahead. Given the choice no child would want to be bored or bewildered and, providing the atmosphere encourages it, children can take a part in the responsibility for making progress.

The second point underlines the need to think of the four aspect of the teacher's role as continuous and not as sequential steps. Thus 'gaining access to children's ideas' goes on all the time and provides feedback about the effect of action taken and so enables adjustments to be made in the light of new information from the child's reactions.

Taking action to help development

Reference has already been made earlier in this chapter to various kinds of action in response to information about children's ideas.

They will be drawn together here under three headings: (1) practical activities (2) discussion and (3) expectations and support. These are ways of meeting individual needs within a group context. They do not assume that separate activities will be given to children individually but that, as happens anyway, children will be adapting and extending their tasks in developing their own understandings.

(1) *Practical activities* designed to develop children's ideas are mainly of two kinds:

- *Those which are intended to extend children's experience.* Children's ideas may be limited by their experience as, for example, when they assume that soil is essential for plants to grow or have little clear idea about how living things are adapted to their environments because they have not made many detailed studies of plants and animals in their environments. In one sense these activities will be challenging views based on very restricted evidence but this is not in any way to suggest that the children's ideas are wrong but more in the spirit of 'find out about all of this and see what you think'.

- *Those which involve children in testing their ideas.* For this the ideas have to be testable (see page 14) and in the form of an explanation, although they may well be expressed in various ways by children. 'Oh, I know, it's . . .' does not sound like a hypothesis but it will contain some sentiment that a particular effect has a certain cause. 'The ice melts because it is outside' is one such statement from a child which at first caused some surprise to his teacher. A prediction from this was that if the ice was somewhere else (inside) then it would not melt. The hypothesis was based in this case on the place where the ice was (which was, after all, where ice had been seen to form and melt) and not on the temperature of the place. The prediction could readily be investigated and thus the idea tested.

  The context in this case was a class discussion of what made ice melt, during which several different ideas were put forward and tested. The children whose ideas were not supported by their investigations therefore had immediate access to another idea which worked in practice better than theirs. Thus the whole experience, of discussion and sharing ideas - and not just the activity itself - is important to the development and change in ideas.

(2) *Discussion of words and other representations* is not sufficient on

its own but as an accompaniment to practical activity it can make all the difference to the thinking which is provoked by experience. In an activity about sound, for instance, children in the early junior years may begin to use the word 'vibration', realising that there is something which is described by this word. It will naturally be used at first for 'vibration' which can be felt or seen and thus related to the use of the word in everyday experience, for the experience of being in contact with things which have motors in them, such as a washing machine, vacuum cleaner , electric mixer or power drill. So it will be easy to use the word in relation to a drum which is beaten or a guitar string which has been plucked. But what do children think is happening in a vibration? What examples of vibrating objects can they give? Do things which are not vibrating make a sound?

Discussion of these things, with the children supplying instances of objects explored in their practical work will help them think again about what they have done (and perhaps go back to check things they took almost for granted) as well as reflect on the words they use. At a later stage and with older children the questions would be more challenging and aimed at taking their notion of vibration further and towards vibrations which cannot be seen, such as those in air. Do things which make a sound vibrate? Can you have a sound without vibration? How does the sound travel from a vibrating object to our ears?

(3) *Expectations and support* can be varied from one child to another so that each is enjoying the challenge of extending ideas but within the range of present capabilities. Adjusting these things for each child is a hidden but yet a most powerful means of matching experiences to individual needs. Within a group the amount of help given to children can be different and a teacher can monitor the progress made by individuals in terms of the support they need.

Similarly there can be different goals for the children in the teacher's mind. For example, in response to realising that six year old Simon's ideas about time needed more practice, his teacher did not separate him from group activities but asked him to find out, for instance, how long his cakes took to cook, how long he took to change his shoes and introduced some tasks related to time as appropriate in other activities. In many cases very different responses to the same activities are accepted and praised because the teacher judges them against the effort made rather than the performance in more objective terms. (See Chapter 20)

Knowing when to stop

For each child there is a gap between present ideas and those which we aim to help him or her achieve. The question of how to bridge this gap is one to which there is no universal answer. If there were, education would be more an exact science than a form of art; we would be able to build a child's learning rather as building a house, knowing just how one brick should be placed on another. As we, know, it is not like that. With children's learning the 'bricks' are not ready formed; they need time to take shape and until that has happened there can be no further building. There are occasions when children's ideas need time before there can be further development. This is the moment to stop and it is as important to recognise this as it is to stimulate thinking at other times.

So when is it time to stop? Children's general reactions will indicate this in science, as in other subject areas. When they become easily distracted after a period of working with full attention on something; when they lose interest and adopt a mechanical rather than a thinking approach to their work. These can be signs that the child is no longer in charge of the learning and things may have got ahead of him or her.

In science the signs are also to be found in what may seem to be stubborn persistence in ideas despite evidence which conflicts. For example, the teacher who challenged children's notion that the sound from outside the room must be coming through air round the sides of the door by putting sticky tape all around and over the key hole, found that the children were not convinced that the sound they could still hear was coming through the wood. They claimed that there must be holes which had not been taped over and were letting the sound through. At that time she could not provide convincing evidence of sound passing through solids and was wise enough to leave the children with their idea. At some stage later, with more experience, the idea could be tested more convincingly and further evidence of sound travelling through solids and liquids could be provided. When children hold onto their ideas in this way it is a sure sign that they are not ready to relinquish present ideas for different ones.

For all of us there are times when looking at things a different way is exciting and seems to bring several things into place. There are also time when we can't see that any idea different from our own makes sense and we need more time or more examples if we are going to change our view. We need only think of how difficult it is to change people's view of what foods are good for them to realise that we all hang on to cherished ideas.

So the time to stop trying to help children advance their ideas is when they do not see other ideas as being as useful as their present ones. The time to stop is also *before* falling to the temptation of pressing different ideas with the force of authority, giving the impression that 'this is how things are'. If children feel obliged to accept different ideas because these are clearly 'right' and their own are 'wrong' then they will quite soon lose confidence in their ability to think things out and come to a useful conclusion. It is far better to leave them with an imperfect notion of how sound travels than to turn science into something which they have to accept but don't understand.

CHAPTER 11

The teacher's role in the development of process skills

The mental and physical skills which are involved when children attempt to use and test their ideas in relation to new experience were described in Chapter 4. The interconnectedness of the skills was pointed out there, a feature which needs to be borne in mind here, too, as we consider how the development of these skills can be encouraged. For this reason we begin with some general points which apply to all the skills and then consider those specific to each of the six areas of skills identified in Chapter 4.

Supporting process skill development

At the general level, the teacher's role in providing children with experiences which help them develop process skills has these five aspects:

* Providing opportunity to use process skills in the exploration of materials and phenomena at first hand. This enables children to use their own senses and to gather evidence from which to raise questions, form hypotheses based on existing ideas, and so on. Children have to *use* process skills in order to develop them. Being told about what it means to observe, interpret or investigate is not the same as doing these things. *Action provides the practical basis for thinking*.
* Providing opportunity for discussion in small groups and as a whole class. Tasks which are designed to require children to share their ideas, to listen to others, to explain and defend their ideas will necesarily involve them in thinking through what they have done, relating ideas to evidence and considering others' ways of approaching a problem in addition to their own (see Chapter 13

for more on this). *Talking and listening provides the thinking basis for action.*

- Listening to their talk and studying their products to find out the processes which have been used in forming their ideas. At all stages of activities the teacher can be picking up information about how children have collected and used evidence. *Helping the development of skills depends on knowing how children are using them.*

- Encouraging critical review of *how* activities have been carried out. During and after completing activities children should discuss how they have carried out parts of or the whole of an investigation and be encouraged to consider alternative courses of action and the extent to which these may be improvements. This will enable the children to recognise the skills which they need to improve. *Helping children to realise the skills they need is important for giving them a part in their own learning.*

- Providing access to the techniques needed for advancing skills. In order to increase the accuracy of observation and measurement, for example, the use of instruments needs to be taught as the need for them arises. Other techniques, such as used in the drawing of charts and graphs, and the knowledge of conventions in diagrams, are required for communication. Knowing how to use these instruments and conventions is not the same as using them appropriately, so there is more to using these skills than the basic knowledge. *However, using techniques appropriately requires knowledge of how to use them.*

Teachers can initiate the use of process skills through the questions they ask and examples of questions which encourage the use of process skills can be found in Chapter 14.

Helping development in observing

Opportunity to use the senses as ways of finding out requires objects and phenomena to explore. An 'interest' or science table in the classroom is one way of providing this as well as serving other purposes. It is always a good idea to set out objects relating to a new topic two or three weeks ahead of starting it in order to create interest. During the topic work items can be added to the display. The table enables children to use odd moments as well as science activity time for observing and so increases an important commodity in the development of this skill. *Time* is significant here more than perhaps for other skills. Children need time to go back to things they may have observed only superficially or when a question has occurred to them that suggests something they want to check.

Some children also need *invitations to observe*. Cards placed next to objects or equipment displayed can encourage observation and action. 'Try to make this bottle make a high and a low sound' placed next to a bottle three-quarters full of water encourages interaction. 'How many different kinds of grass are there here?' placed next to a bunch of dried grasses encourages careful observation. The correct use of a magnifying glass can also be taught through a card with a drawing on it. Older children with the required manipulative skill can learn to use a microscope through similar informal opportunities.

When observations have been made there should be opportunities created for them to be shared. Making a point of spending a few minutes as a whole class discussing what has been noticed about things on the science table, for example, may draw the attention of some children to things they have missed and emphasise the role of the table in the class activities.

Not all observations are made in the classroom, of course, and careful preparation for expeditions outside are important if things are not to be missed. There is less opportunity to revisit objects and so it is essential for the teacher to explore in advance the place to be visited, keeping the capabilities and knowledge of the children in mind. Since the school environment is a resource for all classes, it is best if any particular visit is coordinated with other teachers and set into the planning of the whole school programme for science. More detailed suggestions for using the school environment for science can be found in *Environmental Science in the Primary Curriculum* (Elstgeest and Harlen, 1990).

Helping development in hypothesising

As we saw in Chapter 4, a hypothesis is an attempt to explain some observation, happening or relationship. There are things to avoid as well as to encourage in helping the development of this skill. To be avoided is the idea that a hypothesis has to be 'right', that is, that it depends on knowing all about what is happening. This impression can be conveyed through the questions which are asked of children. If the question is framed as 'Why do some leaves turn brown in autumn?' then it is difficult to answer unless you know, or think you know the reason. On the other hand, such questions as 'Why do you think some leaves turn brown?' or 'What do you think could be the reason for leaves turning brown?' stimulate the generation of an explanation from existing ideas. (See Chapter 14 on the wording of questions).

The development of confidence in ability to suggest explanations is helped by asking for several possible alternatives in situations where there is not necessarily an obvious answer. For example, 'Why are there patches of different coloured grass on the playing field?' There are multiple possible reasons, which could be brought out by asking 'What else could be the reason?' as each suggestion is accepted. Each one has to be possible in terms of the evidence – different seeds used; something in the soil under the patches; drainage varying from one place to another; but a suggestion that more rain falls on the patches might be rejected because of evidence that there is nothing to cause such difference. Trying to explain a shared observation such as this enables children to feel that they have the ability to make sense of the things around them. Which of the suggestions may be most likely would require more evidence and investigation but their ability to attempt explanations is not dependent on the result.

Young children's hypotheses will be in the form of attempts to explain specific events in their experience rather than in terms of broad statements of principles which explain a whole range of phenomena (such as the conditions needed for growth of plants to explain the patches in the grass). However, the ability to suggest explanations in specific terms is the foundation of later development in applying broad principles and theories. Meanwhile the skill of using existing ideas, limited though they may be, in attempting explanations plays an important role in testing and developing these ideas.

Helping development in predicting

Opportunities to make predictions can be created both in relation to patterns found in observations and in relation to hypotheses which are put forward to explain observations.

In the case of *patterns found between two variables* the prediction is based on the evidence of some association between one thing and another, but without necessarily assuming that it is an association of cause and effect. The simple relationship between hand size and foot size is an example of association where there is no sense in suggesting that one thing causes the other, that having large hands causes large feet; rather there is another variable which is causally related to both of them. Nevertheless, the association, whatever its basis, can be used in predicting (within limits) the foot size of someone from their hand size.

Not all patterns are simple and the process of predicting from them is best encouraged through the more obvious relationships, such as the sound that is made when a thin strip of wood (or a ruler) is held at one end and twanged at the other. The note varies with the length which is free to vibrate. Once observations have led to the pattern being established, children can predict whether a particular length will give a higher or lower note and then try it out. In such situations it is useful for them to discuss how they make their predictions since this helps them to become conscious of using the pattern they have noticed in the observations. They should recognise that this is diffferent from a guess.

Predictions based on hypotheses depend on previous experience and ideas derived from it rather than the interpretation of findings from an investigation or observed events. Hypotheses do suggest explanations in terms of cause and effect and constitute some theory of why things are as they are. It is the test of a true theory that it can be used predictively. For example the hypothesis that the leaves on some trees turn brown 'because of the cold' is a theory that cold brings about the change. Although not quite accurate this theory can be used to make a prediction about the circumstances in which leaves will turn brown, which is a test of the hypothesis and the idea on which it is based. Again, it is important for children to be helped to make predictions in simple cases and think about the way in which they arrive at a prediction.

Often there seems to be little difference between the two kinds of prediction; it depends on the intention to describe or explain. For example, the distance which a wind up toy car will travel will depend on the number of turns given to the winder. This relationship can be used to predict how far the car will go for a certain number of turns without any attempt to suggest why the distance will change with the winding. But there could be a hypothesis that the distance depends on the energy stored in the spring and so the more energy the further the car will go. The test of the prediction in the second case would be a test of the theory or idea that more energy makes the car go further, in contrast with the first case where there is no theory of cause and effect being tested.

Children often implicitly use patterns or hypotheses in making predictions but fail to recognise that they in fact do so. A girl who had investigated how far a wind up toy travelled after different numbers of turns, after making a prediction of how far it would go for a certain number of turns, said that she had guessed. Further probing, however, led to her describing that 'I thought that it would be a bit more than for 3 and a bit less than for 5', suggesting that she was implicitly using a relationship of 'more winds means further'. Becoming aware of the pattern she was using enabled her to predict other distances with more confidence and indeed satisfaction. Discussion played a central part in bringing about this awareness.

Helping development in investigating

As in Chapter 4 the process of investigating is taken here to be what happens between having a question to investigate or a prediction to test and obtaining observations or data to be interpreted. Too often children's experience of these steps in answering quetions is of following instructions, as in the parachute activity on p65 (Chapter 8) or when teachers guide activities too strongly as in the following classroom observation of a teacher introducing an activity to find out if ice melts more quickly in air or in water at room temperature:

> You'll need to use the same sized ice-cubes. Make sure you have everything ready before you take the ice cubes out of the tray. Put one cube in the water and one close to it in the air. Then start the clock . . .

Here the children will have no problem in doing what is required, but they may have no idea of why they are doing it. If they did, they might challenge the need for a clock in this activity!

A diet weighted with such activities does not give children opportunity to carry out planning and thinking about what they are doing. Following someone else's plans is not the same as planning and moreover it effectively discourages thinking as the activity goes on. The reaction to any problem will be 'this doesn't work', laying the blame outside themselves, rather then being in control of the investigation and taking responsibility for overcoming problems. There must, therefore, be opportunities for children to start from a question for investigation and to think out and carry out their own procedures for answering it. This is asking a great deal from young children and from older ones unused to devising investigations and they will need help which subsequently can gradually be withdrawn.

Young children's experience should include simple problems such that they can easily respond to 'How will you do this?' For example, 'How can you find out if the light from the torch will shine through this fabric, this piece of plastic, this jar of water, this coat sleeve?' Often young children will respond by showing rather than describing what to do. With greater experience and ability to 'think through actions' before doing them they can be encouraged to think ahead more and more, which is one of the values of planning. Involving children in planning is part of setting an expectation that they will think through what they are going to do as far as possible.

For older children, help in planning can begin, paradoxically, from reviewing an investiagtion which has been completed (whether or not the children planned it themselves), helping them to go through what was done and identifying the structure of the activiy through questions such as

- what were they trying to find out?
- what things did they compare? (identifying the independent variable)
- how did they make sure that it was fair? (identifying the variables which should be kept the same)
- how did they find the result? (identifying the dependent variable).

When planning a new investigation the lessons learned from reviewing can be recalled, where perhaps variables were not controlled or initial observations taken when they should have been. Planning continues throughout an investigation and indeed the initial plan may change as the work progresses and unforeseen practical obstacles emerge. However it is important for children to recognise when they do change plans and to review the whole planning framework when a change is made. Writing plans down is a useful activity because it requires forward thinking, actions carried out in the mind. Children become more able to do this the more experience they have to think through and call upon in anticipating the results of certain actions. The teacher's role is thus to provide time and a structure for planning and gradually to set the expectation that children think through what they do even it they do not write formal plans on paper for every investigation.

Helping development in interpreting and drawing conclusions

For children to develop ideas as a result of collecting information and evidence to test their ideas they have to interpret what they find. That is, they must go further than collecting individual observations to see patterns, relate various pieces of information to each other and to ideas. For example, when measuring the length of the shadow of a stick placed in the ground at different times of the day children must go beyond just collecting the measurements if the activity is to have value for developing ideas. The pattern of decreasing and then increasing length of the shadow and the possibility of using the pattern to make predictions about the length at times not measured, or the time of day from the measurement of the shadow, and the development of ideas about how shadows are formed, are important outcomes from this activity. They all depend on *using* the results the children obtain, so the development of the skills required is important. The central part of the teacher's role is to ensure that results *are* used and children don't rush from one activity to another without talking about and thinking through what their results mean.

It was noted earlier (p. 88) that children often make predictions on the basis of patterns without apparently being conscious of doing so. Teachers can help to foster this consciousness by discussing simple patterns, such as the relationship between the position of the sun and the length of the shadow (or the equivalent in a classroom simulation using a torch and a stick). The starting point must be the various ways in which children wil express their conclusions, including:

— 'the shadow is shortest when the sun is highest'
— 'the shortest one is when the sun is high and the longest when it is low'
— 'its length depends on where the sun is'

and working towards recognising that

'the higher the sun the shorter the shadow'

says all that the previous statements say and more besides. Time for discussion of how to express a pattern is essential for the development of this skill.

In this particular example an explanation for the relationship can be found. By asking 'Why do you think the shadow changes length when the sun is higher?' the children's ideas about this can be collected. Other work with shadows can be called upon to bring out the hypothesis that it is because the light is cut off by the stick and carries on in a straight line. Such a 'conclusion', and any others the children may prefer, is no more than a hypothesis to be tested by further evidence. A delicate touch is required to help children work towards a conclusion but still realise that there is always the possibility of evidence being found which does not fit (see Stephen Hawking's comment quoted on page 5).

Helping development in communicating

In the course of their science activities there is the potential for children to experience a range of different kinds of communication for different purposes and audiences. These should include:

Modes:
Writing, speaking, drawing, making, keeping notes, listening, reading, looking.
Audiences:
For themselves, other pupils, their teacher, other adults.
Purposes:
In order to sort out ideas, tell others about what they have done, present observations, findings and conclusions.

This is wide range to cover and clearly not all will be part of every activity. It is useful, however, for a teacher to plan this part of children's activities so that all are included appropriately and regularly. 'Appropriately' means that it should serve the purposes of the activity and not become meaningless ritual. The routine 'write about what you did' can kill any creativity in communication, as well as being a deterrent for some children, like the boy who dreaded a class visit to the museum, even though he loved going there, because it would inevitably be followed by the request for the kind of writing he did not enjoy.

In this chapter we shall confine discussion to communication on paper and through artefacts, since the whole of Chapter 13 is about speaking and listening. Conflating audience and purpose, we look at informal and formal communication. Both are important in children's learning for writing is not to be seen as the means for finding if learning has taken place but an integral part of that learning.

Keeping notes and records during activities

A note-book which is really your own, a private place to write reminders to yourself, notes of various kinds, is a very useful thing. At the least it is an aid to memory and at best a means of having a dialogue with oneself and it assists reflective thinking. It is useful for children to use a note-book, too, to help them to organise their thinking, write rough plans, record observations. It will be a place where drawing and diagrams may take as important a part as words and where words don't have to be marshalled into sentences. In many classes, however, writing informally for themselves in this way is unfamiliar and almost all that children write is formal. There is considerable value in children using a note-book, but they need help. The help has to be given very subtly, though; if there is too much checking-up on what is written, the note-book becomes just another exercise book which is 'marked' by the teacher. The kind of help which is likely to be effective when note-books are first being used will include:

(a) Opportunity – a suitable note-book and time to use it.
(b) Suggestions for how to use it – when setting the scene for activities and explaining the organisation of the work, include comments on what it would be useful to note down (these should come from the children as well as the teacher)
(c) Help in recording different kinds of information – give ideas for drawing diagrams so that essentials only are recorded, for labelling and annotating drawings, for tabulating information.
(d) Occasional and casual discussion of how the note-book is being used – make non-judgemental comments and give helpful suggestions as in (c).
(e) Showing an example – use a note-book yourself, particularly on trips out of the classroom to note points to discuss later.

Children should begin using notebooks as soon as writing becomes fluent. It is probably best to introduce them to the whole class, encouraging those less able to write to draw and use what words they can.

Making a formal record

The form that a formal record of activities takes should be varied and discussed as part of the scene setting for an activity. Often it will be a product of group effort and will be intended for display in the classroom. Discussion, either with a group or with the whole class if all have been doing similar things, of the best way to present information is the opportunity to introduce techniques for graphical representation or, more often, to talk about how to select the best way of presenting information. Work already displayed on the wall can be used as examples of how to and how not to do this. Children are usually willing to criticise their own work after some time has elapsed. It is also a good idea to have one or two examples of commercial posters (such as those about the nutritional value of different foods or showing types of clouds) to show different ways of providing information.

Looking at posters and books is the other side of formal communication, that is, using secondary sources of information. Children need opportunity – suitable reference sources and time to use them – and some help in locating and selecting information.

In summary the teacher's role is to

- Conduct discussions of ways of communicating particular information to particular audiences.
- Introduce techniques for presenting information, through direct teaching of conventions and providing examples.
- Make suitable reference books available.
- Encourage critical discussion of their own and others' ways of recording and presenting results.

CHAPTER 12

The teacher's role in developing scientific attitudes

In this chapter we look at the teacher's role in development of attitudes of the two kinds discussed in Chapter 5. Attitudes towards the self in relation to school work affect all work and so must be considered in relation to learning science. In addition, the scientific attitudes which were identified as attitudes *of* science (as distinct from attitudes towards science), have an important role in determining the way in which scientific activity is carried out.

It is useful to start with the reminder that attitudes are generalised aspects of behaviour which cannot be taught in the sense of giving instruction. Rather they are 'caught', picked up through example and selectively encouraged through praise and approval. Hence the teacher's role is particularly crucial and there are both things to avoid in this role and positive actions to take.

Encouraging positive attitudes in relation to school work

The aim of action in relation to these attitudes is to avoid the development of the vicious circle described in Chapter 5, whereby children see themselves as failing even before they begin a task and therefore make little effort, leading to failure which confirms their prophecy. One important action to be avoided is the *labelling* of children either as groups or individuals. When asked how she dealt with pupils working at different rates a teacher explained:

> Well, they are the B stream, you see. I mean, I know that there are bound to be some who are better than others, but they all get on with the same work and they enjoy it.
>
> (From *Match and Mismatch: Raising Questions*, 1977)

It is difficult to imagine that the teacher's view of them as 'the B stream', reinforced by the uniformly low level of work expected, did not transfer to the children's self image.

Teachers recognise that even young children are very alert to the signals around them and have well formed views about the work they are given, about their teacher and those around them. For example, eight year old Christopher described exactly why he found school 'boring':

> When the book says 'write the answer', I have to write the whole sum because Mrs. X says it ... writing . . . but it takes so long. It's ... ... the board until all the class knows it. She just goes on and on ... ... know it's wrong, you cross it out and ... again. Then you have to copy it out because it's messy. You have to copy out all the work, not just the bit where you made a mistake. When she ... bit of my writing, she says 'Look at that i. It's not like ... it's ... little ...an'. She tells all the class and they laugh.
>
> (ibid)

Christopher explained, too, that the reason why he was always accused of ... was because it was more fun than doing the work ... Now it may be that Mrs. X was unaware of Christopher's reaction to certain aspects of her teaching and that, had she known she would have been more sympathetic to him.

So, given that children do have clear ideas about their work, *one important thing that a teacher can do is to find out what these are.* Just showing interest in how the children feel about their work is in itself significant in signalling the importance the teacher attaches to providing work which children will put effort into. Christopher's remarks were made to a sympathetic outsider to the school and it may be difficult for the teacher to obtain such frank statements in discussion with an individual, although this may be possible in some cases. The idea of a regular 'review' involving pupils in discussions with teachers about their work, proposed as part of some schemes of records of achievement at the secondary level, has been suggested at the primary level (Conner, 1991). For older primary children it may also be possible to obtain written comments. The following form by Roger Pols (reproduced in Hopkins, 1985) could be modified by adding requests for short supplementary answers, such as 'which parts (of the lesson) did you enjoy?' to give more help in suggesting action for the teacher to consider.

Information of this kind also helps in working towards a second

A Self-assesment Questionnaire

Please put a ring round the answer you wish to give to each question. If you are not sure ring the nearest to what you think.

1. How much of the lesson did you enjoy?   All of it(Some of it/None
2. How much do you think you learnt?   Nothing/Something/A lot
3. How much did you understand?   Most of it/Some of it/Nothing
4. Could you find the books, information,   None/Some of it/Most of it
   equipment you needed?
5. Did other people help you?   A lot/A little/Not at all
6. Did other people stop you working?   A lot/Sometimes/Not at all
7. Did the teacher help you   Enough/Not enough
8. Did the lesson last   Long enough/Too long/Not long
                                               enough
9. Was the lesson   Boring/Interesting
10. Did you need anything you could not   Yes/No
    find?
11. Where did you get help from?   Teacher/Group/Someone else
12. Did you find this work   Easy/Hard/Just about right
13. Write down anything which made it hard for you to learn
14. Write down anything you particularly
    enjoyed about this lesson

(Quoted in Conner, 1991, p. 93)

Figure 12.1

aim in the development of positive attitudes which is *to encourage children to share responsibility for their learning*. For this they have to be aware of what they are expected to achieve in their work.

It may well be a good use of learning time to spend some of it discussing different ways of going about a task and what the children think would be the best way to tackle something and why. A class of nine year olds readily came up with the home truth (about a project on different countries) that 'if we just copy the book, we don't really understand it' and proposed that they should read first and then put down what they thought. The teacher added to this, 'Yes, then you will also be more careful to understand what you read in a book and that will help you whenever you use books to find information'. It is not easy for anyone to stand back from specific learning to examine the *process* of learning itself and young children are not often able to do this, but a gentle move in this direction can help them realise the point of what they are doing.

Interest in children's feelings and views on their learning has to be sincere. Children are not taken in by the superficial interest of their teacher, for it will be betrayed by manner and tone of voice as well as by whether anything happens as a result. A genuine interest creates an atmosphere in which children's own ideas are encouraged and taken as a starting point, where effort is praised rather than only achievement, where value is attached to each child's endeavours. In this atmosphere, a child who does not achieve as well as others will not be ridiculed. The range of activities available makes allowances for differences in ability of children and the teacher's interest and approach results in involvement of children in their work and their own learning.

Encouraging scientific attitudes

In Chapter 5 we considered four attitudes as being particularly relevant to learning science: curiosity, respect for evidence, willingness to change ideas and critical reflection. As was evident in the previous discussion, these are closely related to each other and this applies even more to the actions which a teacher can take to encourage them. Thus we consider them here as a group to avoid repetition.

The aspects of the teacher's role are of four main kinds:

- showing an example
- reinforcing positive attitudes with praise and approval
- providing opportunity for the development of the attitudes
- discussing situations in which different attitudes would lead to different courses of action.

Showing an example

Given that attitudes are 'caught' this is probably the most important of the positive things that teachers can do. For example, to make a point of revealing that his or her own ideas have changed can have a significant impact on children's willingness to change their ideas; for instance:

> I used to think that trees died after dropping their leaves, until . . .
> I didn't realise that there were different kinds of woodlice . . .

> I thought that it was easier to float in deep water than in shallow water but the investigations showed that it didn't make any difference.

The old adage that 'actions speak louder than words' means that such comments will not be convincing by themselves. It is important for the teacher to show all of the indications of attitudes which were mentioned in Chapter 5, perhaps the most significant of which are

- showing interest in new things (which the children have brought in for example)
- helping to find out about new or unusual things
- admitting when evidence gathered does not seem to fit in with expectations
- suggesting that further evidence is needed before a conclusion is reached
- acknowledging that evidence means a change in ideas
- being self-critical about how things have been done or ideas applied
- admitting when the explanation for something is not known.

In a classroom where useful ideas are pursued as they arise and activities extend beyond well beaten tracks, there are bound to be opportunities for these teacher behaviours to be displayed. Situations in which the teacher just doesn't know, or which bring surprises or something completely new, should be looked upon, not as problems, but as opportunities for transmitting attitudes through example.

Reinforcing positive attitudes through selective approval

Children pick up attitudes not only from example but from what in their own behaviour earns approval or disapproval. When children show indications of positive attitudes, it is important to reinforce these behaviours by praise or other signs of approval. This is far more effective than discouraging negative attitudes. Those who have not developed positive attitudes will be able to recognise what these are from the approval given to others.

For example, if critical reflection leads to children realising that they did not make fair comparisons in their experiment, the teacher's reaction could be 'Well you should have thought of that before' or 'You've learned something important about this kind of investigation – well done'. The latter is clearly more likely to encourage reflection and the admission of fault on future occasions. Moreover if this approval is consistent it eventually becomes part of the classroom climate and children will begin to reinforce the attitudes for themselves and for each other.

Providing opportunity

Since attitudes show in willingness to act in certain ways, there has to be opportunity for children to have the choice of doing so. If their actions are closely controlled by rules or highly structured lesson procedures, then there is little opportunity to develop and show certain attitudes (except perhaps conformity). Providing new and unusual objects in the classroom gives children opportunity to show and satisfy – and so develop – curiosity. Discussing investigations during their course or after they have been completed gives encouragement to reflect critically, but unless such occasions are provided the attitudes cannot be fostered.

Discussing attitude-related behaviour

Attitudes can only be said to exist when they are aspects of a wide range of behaviour. In this regard they are highly abstract and intangible. Identifying them involves a degree of abstract thinking which makes them difficult to discuss, particularly with young children. However, as children become more mature they are more able to reflect on their own behaviour and motivations. It then becomes possible to discuss examples of attitudes in action and to help them identify explicitly the way they affect behaviour.

When some ten year olds read in a book that snails eat strawberries, they tested this out and came the the conclusion that 'as far as our snails are concerned, the book is wrong'. Their teacher discussed with them how the author of the book might have come to a different conclusion from them and whether both the author and the children might gather more evidence before arriving at their conclusions. The children not only recognised that what was concluded depended on the attitudes to evidence but also that the conclusions were open to challenge from further evidence, thus developing their own 'respect for evidence'.

CHAPTER 13

Talking, listening and using words in science

This chapter is mainly concerned with spoken language and with oral transactions in the classroom, complementing what has been said about communication on paper in Chapter 11. We will also consider the matter of words used in the talk and the uncertainty which is often found about when and whether to use and expect children to use 'correct' scientific words.

It is useful to draw a distinction, following the ideas of Douglas Barnes, between speech as communication and speech as reflection. Already many references have been made in this book to the value of children discussing with each other, exchanging ideas and developing their own views through the act of trying to express them and explain them to others. This involves both communication and reflection. The reflective part is sorting out their own ideas aloud, indeed 'thinking aloud'. The communication is sharing with others and involves listening and responding in a way which is coherent and understandable by others. Barnes claims that both are needed and that it does not serve learning to focus only on the more formal communication since

> if a teacher is too concerned for neat well-shaped utterances from pupils this may discourage the thinking aloud.
>
> (Barnes, 1976, p. 28)

Speech as reflection

We have all probably had the experience where talking to someone has resulted in developing our own understanding, although apparently nothing was taken from the other person in terms of ideas. The effect is even more striking when you are the one against whom ideas were 'bounced' and are thanked for help in sorting out ideas when all you have done is listen and perhaps question in a neutral manner.

The presence of one or more other people is essential in these cases, not only to legitimate thinking aloud but for offering the occasional comment or question for clarification – this has the effect of provoking reflection on what we think as we express it.

The same thing happens with children though often less tidily, since the reflection is going on in several minds at the same time. The following example of interaction in a classroom involved the teacher, but acting as one of the group rather than as an authority figure.

Deidre and Allyson were investigating the way in which three whole hens' eggs, labelled A, B and C behaved in tap water and in salty water. They knew that one was hard-boiled, one soft-boiled and one raw. They had to find out which was which. This is how the eggs landed up just after being placed in the salty water. The transcript begins with the teacher approaching them after they had been working alone for some time.

DEIDRE   . . . hard-boiled.
ALLYSON I know. . .
TEACHER (*coming up to them*) Can you tell me how you're getting on?
DEIDRE   I think that C is raw.
ALLYSON We both think that C is raw.
TEACHER Do you?
DEIDRE   B is. . .
TEACHER (*to Allyson*) Why do you think that?
ALLYSON Because when you put eggs in water bad ones rise to the top.
DEIDRE   (*at the same time*) Because it. . .we put them all in. . .
TEACHER Bad?
ALLYSON Yes, I think so—or is it the good ones?. . .well, I don't know.
TEACHER Yes?
ALLYSON . . .they rose to the top, so. . .

Deidre is putting the eggs into the salty water.

DEIDRE   . . .that's the bottom (*pointing to C*).
ALLYSON . . .if it's raw it should stay at the bottom.
TEACHER I see.
DEIDRE   So that's what we think, C is raw and B is medium and A is hard-boiled. (*Allyson starts speaking before she finishes.*)
ALLYSON . . .and I think that B is hard-boiled and she thinks that B is medium.
TEACHER Ah, I see. (*To Deidre*) Can you explain, then, why you think that?

DEIDRE   If we put. . .er. . .take C out, (*takes C out, puts it on the table, and then lifts A and B out*) and put these in, one after the other. Put A in—no, B in first. That's what. . .Allyson thinks that is hard-boiled, I think it's medium. If you put that in. . .(*She puts B into the salty water.*)

ALLYSON   . . .'cos it comes up quicker.

DEIDRE   It come up quick. And if you put that in. . .

She puts A into the salty water. It goes to the bottom and rises very slowly.

ALLYSON   And that one comes up slower.

DEIDRE   So, I think that one (*pointing to A*) is hard-boiled because it's. . .well. . .

ALLYSON   I don't. I think if we work on the principle of that one (*pointing to B*). Then that one comes up quicker because it's—you know, not really boiled. It's still a little bit raw.

TEACHER   A little bit raw.

ALLYSON   So, therefore it'll come up quicker.

DEIDRE   Yes, but it's not bad.

TEACHER   What'll it be like inside?

ALLYSON   Runny.

TEACHER   It'll be runny still. I see.

(*Match & Mismatch: Raising Questions*, 1977)

Having agreed that C is the raw egg, Deidre and Allyson disagree about the identity of the other two eggs. Allyson has a reason for considering B is hard-boiled on the basis that 'bad ones rise to the top', so she considers that B behaves as if it had had something done to it. But she does not articulate the consequences of this until Deidre attempts to give her reason. Then it is as if Deidre's reason, which she interrupts, sparks off her own thinking.

Although they respond to the teacher's request for an explanation what they do is to continue their interaction and struggle to work out their own reasoning. Deidre's response in particular is hesitant and disjointed, not at all like a straight answer to the question. Barnes call this 'exploratory talk' and he argues that

> it is very important whenever we want the learner to take an active part in learning, and to bring what he learns into interaction with that view of the world on which has actions are based. That is, such exploratory talk is one means by which the assimilation and accommodation of new knowledge to the old is carried out.

(Barnes, 1976, p. 28)

We can see from this interchange about the eggs how the girls use evidence to check their ideas. This comes through most clearly in Allyson's 'if we work on the principle that . . . ' where she relates what she predicts on the basis of her judgement to the observation of how quickly the egg floats up in the salty water, but it also occurs throughout. It is worth noting in passing that the origin of her idea is previous knowledge about how to distinguish good from bad eggs.

It is difficult to deny that Deidre and Allyson learn from this experience and from their discussion of it. But this does not always happen. For instance see how June and David, in the absence of the teacher, seem to regard the task as one where giving an answer is more important than having a reason for the answer. June, particularly, seems keen to move to early closure.

DAVID   Look at that one, this one, look, June.
JUNE   That one's the one that's not boiled.
DAVID   How do you know?
JUNE   Oh, I'm not stupid.
DAVID   Shall I put them in there, or in there? (*On the table or in the container where they were first.*)
JUNE   Yea, put them in there.

David puts the one he took out in the container and June brings out the other two eggs.

JUNE   There's B. . .(*as she passes them to David who places them carefully.*) Now put them in salty water. Put them in salty water.

David picks up A and puts it in the jar of salty water.

DAVID   A Floats. A.
JUNE   B (*She puts B in. It sinks.*) Sinks.
DAVID   C.

He puts it in the salty water. It goes to the bottom and slowly begins to rise again.

JUNE   Sinks.
DAVID   Yea look. . .no, it doesn't.
JUNE   No. . .that one (*she points to C. Pauses, uncertain for a moment*). No, how are we going to tell. . .
DAVID   That one's. . .

> JUNE   Hard-boiled. The one at the bottom's hard-boiled. Put C hard-boiled. (*She instructs David to write. But it isn't C which is at the bottom!*)

(ibid)

Even here, though, there are signs that they are close to becoming more involved. David's 'How do you know?' could have sparked June into explaining her ideas, had she been less defensive. Later on, when an egg which she declares 'sinks' begins to float upwards, there is questioning in the air. The potential seems to be there. The question is then how this potential can be exploited; how, more generally, can we encourage interchanges which involve reflective thinking?

> The quality of the discussion – and therefore the quality of the learning – is not determined solely by the ability of the pupils. The nature of the task, their familiarity with the subject matter, their confidence in themselves, their sense of what is expected of them, all these affect the quality of the discussion, and these are all open to influence by the teacher.

(Barnes, 1976, p. 71)

The way in which Deidre and Allyson's teacher acts gives several clues to positive encouragement of reflective thinking:

— Joining in as part of the group, without dominating the discussion.
— Listening to the children's answers and encouraging them to go on ('I see', 'Yes?')
— Asking the children to explain their thinking.
— Probing to clarify meaning ('What'll it be like inside?')

Not all aspects of the teacher's role can be illustrated in one short interchange and indeed much of it consists of setting a context and a classroom climate which encourages exploratory thinking and talk. Important in this respect are:

— Expecting children to explain things, which involves valuing their ideas even if these are unformed and highly conjectural.
— Avoiding giving an impression that only the 'right' answer is acceptable and that children should be making a guess at it.
— Judging the time to intervene and the time when it is better to leave a children-only discussion to proceed.

The presence of the teacher changes a discussion quite dramatically, for it is difficult for him or her not to be seen as an authority. Left alone, children are thrown onto their own thinking and use of evidence. But, as we see with June and David, the absence of a teacher does not always lead to productive interchange and it is not difficult to imagine how a question from a teacher could have supported the move towards enquiry which David seemed to be making. The teacher needs to monitor group discussions, listening in without intervening, before deciding whether 'thinking aloud' is going on usefully or whether it needs to be encouraged.

Speech as communication

This is the more formal side of using talk, where shared conventions and expectations have to be observed if others are to be able to make sense of what is said. It is part of socialization to be able to tell others in a comprehensible way about what has been done or thought about and to be able to listen to others, attending not only to the words but to the implicit messages conveyed in tone of voice and manner. In order to develop necessary skills, children need opportunity for reporting orally to others in a setting where they know that others will be listening and where they have to convey their information clearly. Such situations do not occur naturally in the classroom and have to be set up. The role of the teacher in creating the right climate as well as the specific opportunities is clearly a central one.

How a teacher understands the purpose of getting children together to share experiences and ideas will influence the outcome of such occasions. If it is mainly a social event, with the emphasis on the opportunity to speak and little feedback in relation to the content, then it may become something of a ritual. The teacher may approve anything as long as children 'speak up' and are not long-winded. This does not encourage children to listen carefully to each other and so, knowing that there is not much attention from others, they give slight thought to what they say. On the other hand the teacher who regards what is said as important so that children learn from each other's ideas will indicate this is in comments about the content, comparing and contrasting different contributions, questioning, etc. Then the telling and listening can have a role in the development of children's ideas as well as in their communication skills. It means that they go back over their activities and make sense of them for themselves so that they can make sense of them for others.

In this way, communication will have value in developing both the children's ideas and their communication skills; but particularly if the purpose of discussion is perceived by both teacher and pupils as part of the activity. To further this a teacher needs to

— Make use of children's ideas in comments, thus encouraging children to do the same ('that's an interesting idea you have about . . . ' 'Tell us how you think it explains . . . ')
— Encourage children to respond to each other and not just to make statements of their own ideas.
— Listen attentively and expect the children to do so.
— Set up expectations that children will put effort into their presentations to each other and try to make them interesting; give time and help in preparation with this in mind.

These things have to become part of the general way of working since expectations that children will respond to what their classmates say are set by the pattern of previous lessons as well as by the response on a particular occasion.

Introducing scientific words

Children pick up and use scientific words quite readily; they often enjoy collecting them and trying them out as if they were new possessions. At first one of these words may have a rather 'loose fit' to the idea which it is intended to convey.

Take for example the child's writing in figure 3.10 (p. 25) where, in describing how the sound is transmitted in a yoghourt pot and string telephone, he explains how vibrations go down the string. The word 'vibration' is certainly used in a manner here which suggests that the child understands sound as vibration, until we notice that he writes that the voice is 'transferred into vibrations' at one end and 'transferred back to a voice' at the other. It seems that the sound we hear is not understood as vibration, but only its transmission along the string. It may be that both ideas of sound and of vibration have to be extended, so that vibration is understood as something which can take place in air and occurs wherever sound occurs; this will take time and wider experience, but he has made a start.

The question as to whether the child should be using a word before he can understand its full meaning is a difficult one. But first we have to ask what we mean by 'full meaning'. Most scientific words (such as evaporation, dissolving, power, reflection) label concepts which can be understood at varying levels of complexity. A scientist understands energy in a far broader and more abstract way than the 'person in the street'. Even an apparently simple idea of 'melting' is one which can be grasped in different degrees of complexity: a change which happens to certain substances when they are heated or an increase in energy of molecules to a point which overcomes the binding forces between them. This means that the word 'melting' may evoke quite a different set of ideas and events for one person than for another. Now to use the word 'melting' in a restricted sense is not 'wrong' and we do not insist that it is only used when its full meaning is implied. Indeed the restricted meaning is an essential step to greater elaboration of the concept. Therefore we should accept children's 'loose' use of words as a starting point to development of a more refined and scientific understanding of the word.

Two questions of importance then arise:

- When is it useful to introduce a word, knowing that it can only be used loosely at first?
- How are we to know the meaning that a child has for a word?

We can deal with the second question briefly here. Children will, of course, pick up words without being given them by a teacher. It is important for the teacher to know what a child means by a word, for confusion can be heaped upon confusion if meanings differ. But since finding out children's meaning of words is an integral part of finding out their ideas we will deal with practical suggestions in Chapter 22.

The first question causes much concern. Teachers seem to be caught between, on the one hand giving new words too soon (and so encouraging a verbal facility which conceals misunderstanding) and, on the other, withholding a means of adding precision to thinking and communication (and perhaps leading children to make use of words which are less than helpful).

The value of supplying a word at a particular time will depend on:

- whether or not the child has had experience of the event or phenomenon which it covers
- whether or not the word is needed at that time
- whether or not it is going to add to the child's ability to link related things to each other.

These points apply equally to a teacher supplying the word consciously and overtly as to providing it indirectly by using it. That means that until the moment for introducing the word is right, the teacher should use the language adopted by the children in discussing their experiences. Then, once the word is introduced the teacher should take care to use it correctly. For example if children have been exploring vibrations in a string, a drum skin, tissue paper against a comb, and wanting to talk about what is happening to all these things, it may well be useful to say 'what all these are doing is called "vibrating".' Before this the children and teacher may have called it by descriptive names: trembling, jumping, moving, going up and down.

All this can be summed up by saying that *if a word will fill a gap, a clear need to describe something which has been experienced and is real to the children, then the time is right to introduce it.* With young children one of the conditions for the 'right time' is the physical presence or signs of the phenomenon to which the word refers. Only then can we hope to fit the word to an idea, even loosely. Much more experience of a concept has to follow so that the word becomes attached to the characteristic or property rather than to the actual things present when it was first encountered. But there is no short cut through verbal definitions in abstract terms.

CHAPTER 14

Teachers' questions which help learning

Questions form about one fifth of teachers' utterances in the classroom according to the extensive research by Galton, Simon and Croll (1980). Whether or not this is considered to be excessive, it remains that asking questions is an important dimension of teaching, being a significant variable which characterises teaching style. Teachers ask questions for many purposes – for pupil control, for information, to check up or test recall, to provoke thinking, to prompt and lead in a certain direction, to reveal children's ideas. The concern here is with the content and nature of the questions asked rather than their quantity. We must also confine ourselves to questions which have a particular influence on learning in science.

Ways of categorising questions

Productive and unproductive questions

Some important distinctions between different kinds of question are made by Elstgeest (1985), drawing attention to the timing as well as the content of the questions. He gives the following example of an unproductive and a productive question:

> A child was reflecting sunlight onto the wall with a mirror. The teacher asked: 'Why does the mirror reflect sunlight?' The child had no way of knowing, felt bad about it and learned nothing. Had the teacher asked: 'What do you get when you stand twice as far away from the wall?' the child would have responded by doing just that, and would have seen the answer reflected on the wall.
>
> (Elstgeest, 1985, p. 36)

The teacher's reason for asking this question is not clear and may not have been thought out. Many questions of this kind emerge from what Elstgeest calls the 'testing reflex'. The impulse to ask such questions has to be controlled and the purpose of questions more clearly worked out. Elstgeest dubbed the 'testing' questions as *unproductive* and suggested the following sequence of *productive* questions which roughly correspond with the order in which they could be used to encourage a child's investigation:

(1) *Attention-focusing questions*, which have the purpose of drawing children's attention to features which might otherwise be missed. 'Have you noticed . . .?' 'What do you think of that?' These questions are ones which children often supply for themselves and the teacher may have to raise them only if observation is superficial and attention fleeting.

(2) *Measuring and counting questions* – 'How much?' 'How long?' – take observation into the quantitative and lay a foundation for the next type of question.

(3) *Comparison questions* – 'In what ways are these leaves different?' 'What is the same about these two pieces of rock?' – draw attention to patterns and lay the foundation for using keys and categorising objects and events.

(4) *Action questions* – 'What happens if you shine light from a torch onto a worm?' 'What happens when you put an ice cube into warm water?' 'What happens if . . . ' – are the kinds of question which lead to investigations. There are also some useful action questions suggested in Harlen and Jelly (1989, p. 22).

(5) *Problem-posing questions* – give children a challenge and leave them to work out how to meet it. Questions such as 'Can you find a way to make your string telephone sound clearer?' 'How can you make a coloured shadow?' require children to have experience or knowledge which they can apply in tackling them. Without such knowledge the question may not even make sense to the children. It is in relation to this point that the matter of the right time for a question arises.

Elstgeest uses this anecdote to illustrate the importance of choosing the right moment for a particular type of question:

> I once asked a class of children, 'Can you make your plant grow sideways?' For a short time they had been studying plants growing in tins, pots, boxes and other contraptions made of plastic bags. I was just a little too anxious and too hasty and, quite rightly, I got the answer, 'No we can't'. So we patiently continued with scores of 'what happens if . . . ' experiments. Plants were placed in wet and dry conditions, in dark and in light corners, in big boxes and in cupboards,

inside collars of white and black paper, upside down, on their side, and in various combinations of these. In other words, the children really made it 'difficult and confusing' for the plants. Their plants, however, never failed to respond in one way or another, and slowly the children began to realise that there was relationship between the plant and its environment which they controlled. Noticing the ways in which the plants responded, the children became aware that they could somehow control the the growth of plants in certain ways . . . When the question 'Can you find a way to make your plant grow sideways?' reappeared later there was not only a confident reaction, there was also a good variety of attempts, all sensible, all based on newly acquired experience, and all original.

(Elstgeest, 1985, p. 39–40)

Open and closed questions

Although the distinction between open and closed questions has become familiar there are many occasions where closed questions are asked whilst open ones would be more appropriate. The research of Galton, Simon and Croll (1980) showed that on average only 5% of all teachers' questions were open, whilst almost 20% were closed and 30% requesting facts. Therefore it is perhaps the value of open questions which needs to be pointed out rather than their nature. Open questions give access to children's views about things, their feelings and their ideas, and promote enquiry by the children. Closed questions, whilst still inviting thought about the learning task, require the child to respond to ideas or comments of the teacher. For example these questions:

> 'What do you notice about these crystals?'
> 'What has happened to your bean since you planted it?'

are more likely to lead to answers useful to both teacher and pupils than their closed versions:

> 'Are all the crystals the same size?'
> 'How much has your bean grown since you planted it?'

Closed questions suggest that there is a right answer and children may not attempt an answer if they are afraid of being wrong.

Person-centred and subject-centred questions

Another way of avoiding the 'right answer' deterrent is to recognise the difference between a subject-centred question, which asks directly about the subject matter, and a person-centred question which asks for the child's ideas about the subject matter. Subject-centred questions are such as:

'Why do heavy lorries take longer to stop than lighter ones?'
'Why did your plant grow more quickly in the cupboard?'

which cannot be answered unless you know, or at least think you know, the reasons. By contrast the person-centred versions:

'Why do you think heavy lorries take longer to stop than lighter ones?'
'Why do you think your plant grew more quickly when it was in the cupboard?'

can be attempted by anyone who has been thinking about these things (and we assume the questions would not be asked unless this was the case) and has some ideas about them, whether or not correct. Where there is interest in children's reasoning, person-centred questions are clearly essential, but at other times they are a more effective, and more friendly, way of involving children in discussions which help in making sense of their work.

Questions to promote thinking and action

We shall be considering here questions which are designed for three purposes: finding out about children's ideas, developing their ideas and leading to the use and development of process skills. The kinds of questions which seem best suited to these purposes can be described in terms of the categories above.

Questions for finding out children's ideas

The following questions were among those designed by the SPACE project to be used by teachers to find out children's ideas. These particular question were used when children had been involved in handling, observing and drawing sprouting and non-sprouting potatoes:

'What do you think is coming out of the potato?'
'What do you think is happening inside the potato?'
'Why do you think this is happening to the potato?'
'Do you think the potato plant will go on growing?'
'Can you think of anything else that this happens to?'

(Russell and Watt, 1990, p. A-10)

They can readily be seen to be open, person-centred questions, since there was a need for children to be given every encouragement to express their thoughts at that time, before investigations started, so that the teachers would know the children's initial ideas.

Questions for developing children's ideas

According to the kinds of ideas the children start from, activities to develop them may take various forms: testing ideas by using them to make predictions and then devising an investigation to see if there is evidence of the prediction being correct, applying ideas in problem-solving, making further observations or comparisons, discussing the meaning of words, consulting secondary sources. Questions can be used to initiate these activities and children's participation in planning them.

Encouraging children to test their ideas means that these ideas have first of all to be in a testable form. 'It's the rudder which makes this boat go better than before' is not testable until the particular aspects of the rudder and the meaning of 'go better' are specified. Questions of the kind:

'How would you show that your idea works?'
'What would happen which showed that it was better?'
'What could you do to make it even better?'

require the specification of variables which are only vaguely identified in the initial statement. When it comes to carrying out a test then many of the questions for developing process skills, in the next section, could be used.

When the development of children's ideas seems to require further experience and comparisons between things, then attention-focusing, measuring and counting and comparison questions are the most useful. For applying ideas, the problem-posing questions are appropriate. For discussing meaning of words, it is best to ask for examples rather than abstract definitions. One of the words which children often use imprecisely is 'dissolving'; indeed they often talk about their own actions as 'dissolving' things, when they probably mean mixing together. These uses can be clarified through questions such as:

'Show me what you do to 'dissolve' the butter?'
'What will happen to the sugar if it dissolves?'
'How can you make something dissolve?'

and other questions for clarifying words discussed in Chapter 13.

Questions for developing process skills

Here we bring together a list of question which encourage children in using the six process skills discussed in Chapters 4 and 11. The content is derived from an activity with seeds and soil involving the use of several types of seeds and several of each one, magnifying lenses, soil, water and pots in which to plant the seeds.

Observing

- What do you notice that is the same about these seeds?
- What differences do you notice between seeds of the same kind?
- Could you tell the difference between them with your eyes closed?
- What happens when you look at them using the lens?

Hypothesizing

- Why do you think the seeds are not growing now?
- What do you think will make them grow faster?
- Why would that make them grow faster?
- Why do you think the soil helps them to grow?

and later when seeds have been planted and are growing –

- Why do you think these are growing taller than those?
- What do you think has happened to the seed?
- Where do you think these leaves have come from?

Predicting

- What do you think the seeds will grow into?
- What can we do to them to make them grow faster?
- What do you think will happen if they aren't in soil but get some water in another way?

and, in relation to growing plants –

- What do you think will happen if we give them more (or less) water/light/warmth?

Investigating
- What will you need to do to find out . . . (if the seeds need soil to grow)?
- How will you make it fair (i.e. make sure that it is the soil and not something else which is making the seed grow)?
- What equipment will you need?

- What will you look for to find out the result?

Interpreting findings and drawing conclusions

- Did you find any connection between . . . (how fast the plant grew and the amount of water/light/warmth it had?)
- Is there any connection between the size of the seed planted and the size of the plant?
- What did make a difference to how fast the seeds began to grow?
- Was soil necessary for the seeds to grow?

Communicating

- How are you going to keep a record of what you did in the investigation and what happened?
- How can you explain to the others what you did and found?
- What kind of a chart/graph/drawing would be the best way to show the results?

CHAPTER 15

Encouraging children's questions

Before embarking on the discussion of how to encourage children's questions we should first establish why it is important for children to ask questions and what kinds of questions we want to encourage. The matter of what to do in response to questions raised as a result of successful encouragement is taken up in Chapter 16.

Why encourage questioning?

Asking questions is central to exploring and trying to understand one's environment. When engrossed in the study of something new we use our existing knowledge to make sense of it and try out the ideas we already have to see if they fit. When we find a gap between what we already know and making sense of something new, one way of trying to bridge it is to ask questions. We might do this immediately by asking a question if there is an authority present, as might happen at an exhibition, on a guided tour, or in a class or lecture. At other times the question may remain unspoken but guides us to a source of information which is then more efficiently used because we know what ideas or information we are looking for.

The relationship of asking questions to curiosity was discussed in Chapter 5. Just as a development from an immature to a mature curiosity can be identified, there is also development in the ability to pose questions which help understanding. Superficial questions, with little interest in the answer, are symptomatic of immaturity in this kind of questioning. There is some development when a child makes it clear that something is not making sense. The child is not accepting things uncritically but is not able to express the query in the form of an answerable question. The more mature ability is shown by clear and often detailed questions which make clear the nature of the gap between understanding and information. The child who can identify in this way the cutting edge of his or her learning is taking an important part in that learning. Thus is it very important for teachers to encourage the development of this ability.

Although we are in the main concerned with questions expressed verbally, it is the case that questions at all levels of maturity can be expressed in other ways. Indeed it is through action that very young children express curiosity.

> Tiny Niels, beaming, bare and beautiful, crawled on the wet sand on the beach. He moved where the sea reaches out for the land, where the ocean barely touches the continent, where the exhausted waves drag themselves up the incline and withdraw or sink into the sand. Whenever this happened in slow and steady rhythm there appeared, all around Niels, tiny holes in the sand which bubbled and boiled with escaping air. These little marvels drew his attention, and with immense concentration he poked his finger in hole after hole . . . The bubbling holes invited Niels: 'Come here, look at us, feel and poke.' And Niels did exactly that. He could not talk yet, not a word was exchanged, no question formulated, but the boy himself was the questions, a living query: 'What is this? What does it do? How does it feel?'
>
> (Jos Elstgeest, 1985, p. 9)

It happens also with older children when they need to take more planned action to answer their questions:

> Some junior boys were using three sand-timers of different durations. During their activity they found that turning over the one-minute timer five times did not take as long as the five-minute timer. They then checked the one-minute timer against the classroom wall clock, which had a second hand. As there was a difference there was then a problem of deciding which was correct. The question in their minds was clear from their action, of going into other rooms in the building to check the timer against different clocks.
>
> (*Match and Mismatch: Raising Questions*, 1977)

Which kinds of question?

We all, adults and children alike, ask a number of kinds of question apart from those seeking information or ideas. Some questions are rhetorical and some just show interest; neither of these expects an answer. Some questions are asked to establish a relationship with someone, or to gain a response; some to attract attention; some even to irritate or harass (as in Parliaments). Questions which arise from curiosity and the desire to understand have the main part to play in learning science but it is important not to discourage any questions by implying that only some are worth answering.

Within the 'desire to understand' questions it is necessary to recall (from Chapter 1) that science is able to answer only certain kinds. So whilst we recognise the value to children of encouraging the expression of their questions, including the vague and unspoken ones, it is helpful to their learning if they begin to recognise the kinds of questions which can be addressed through scientific activity.

Science is concerned with questions about the 'what, how and why' of objects and relationships in the physical world. The most productive kind from the point of view of *learning* science are those which enable children to realise that they can raise and answer questions for themselves. These are the questions which keep alive the close interaction (as between Niels and the holes in the sand) between child and environment, between question and answer.

Children who realise that they can find out answers to 'what, how and why' questions by their own interaction with things around have made the best start they can in scientific development. They realise that the answer to 'why do daisies spread out their leaves? why do paper tissues have three thin layers rather than one thick one? what happens when you turn a mirror upsidedown?' are to be found by directing the questions to the daisies, the tissues, the mirror. 'Asking the object' (a phrase originated by Jos Elstgeest) is the entry to scientific investigation, for it directs the question to where the answer can be found.

Encouraging children to ask questions

Active encouragement is necessary for such an important aspect of children's learning; it cannot be left to chance. Three useful and tested ways of promoting questions generally and scientifically investigable questions in particular are described here.

Providing a classroom climate which invites questions

The close dependence of questioning on curiosity means that the classroom where questions are stimulated must be one where there are plenty of opportunities for direct exploration of interesting materials. Materials and objects brought in by children have built-in interest at least for those who collected them and sharing this interest is very likely to spread it to others. These materials and those brought in by the teacher with a particular view to creating interest in a topic to be studied, need to be displayed in a way which makes it obvious to the children that they are invited to touch, smell (where appropriate and safe), look carefully, find out, etc. In the same vein the expectation that children will ask question should be build up. Sheila Jelly has the following suggestions for ways of encouraging questions, through displays and other tactics:

1. By making sure that displays and collections have associated enquiry questions for the children to read, ponder and perhaps explore incidentally to the main work of the class.
2. By introducing a problem corner or a 'question of the week' activity where materials and associated questions are on offer to the children as a stimulus to thought and action which might be incorporated into classwork.
3. By making 'questions to investigate' lists that can be linked to popular information books.
4. By ensuring that in any teacher-made work cards there is a question framed to encourage children to see their work as enquiry-based and which also provides a useful heading for any resultant work displayed in the classroom.

(Sheila Jelly, 1985, p. 51)

Positive reinforcement

Although it is the new or unusual which is normally a stimulus to curiosity, more familiar objects may be more productive in encouraging children to express questions for investigation, perhaps because they are likely already to have queries in their minds which can be released by the invitation to express them. A display of different tools, nuts and bolts and screws could be set up with a 'question box' for children to post their questions on small pieces of paper. The apparently gender-biased subject matter produced no bias in the interest and questions when this was put into practice. When the box was opened and each question considered, girls were as ready as boys to come up with reasons for different sizes and shapes of heads of screws, why screws had threads but nails did not or whether the length of the handle of a screw driver made any difference. They followed some suggestions up through practical investigations and others were left pinned to the display board awaiting information from an 'expert'. The work added considerably to their experience of

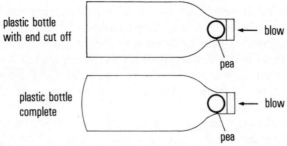

Figure 15.1

materials and their properties as well as showing that questions were valued.

Another way of stimulating questions without constant recourse to things which the children will not have seen before, is to draw attention to important aspects by putting things together with very different properties. For instance, when investigating bouncing balls, what do children think of how a ball of plasticine 'bounces'? Questioning why the plasticine becomes flattened gives an important clue to why balls bounce back. Children can also be made to think about the presence of air in an 'empty' bottle by questioning why they can blow the pea into one of the bottles in Figure 15.1 but not into the other.

Setting aside time for children to describe what they have done is an important part of science experiences which can be used to encourage questions and make it legitimate for children to express questions and admit that there are things they don't know but want to know. It is best for the questions to come from the children rather than the teacher. So instead of the teacher interrogating the children, they should be invited to respond to 'What do you still want to know about . . .?'

More generally, the simple request 'What questions would you like to ask about . . .?' can be regularly extended, either orally or in writing, on work cards or sheets. Resisting the temptation, as a teacher, to do all the question raising is also a simple but important guideline. Raising questions is something children must learn to do for themselves, and this won't be encouraged if all the questions they pursue are raised for them.

Clarifying children's questions

Reference was made in Chapter 6 to the 'Interactive' approach to learning which is based on encouraging children to ask and then investigate questions. Questions are first gathered in various ways similar to those just described but it is recognised that the questions are not always clear and even more infrequently expressed as investigable questions. The following examples are given of teachers helping children to clarify their questions:

About snail shells

Mere: Why do snails have four rings on their shell?

Teacher: Oh . . . do they? Have you checked that out?

Mere: No . . . we checked the two we had.

          Teacher: Would you mind if we changed the question . . .?
          How might you change it?

Mere: What about 'Do snails . . . the garden sort . . . have the same number of rings on their shells?

About fish swimming

Teacher: Can I ask you what you mean by your question 'How does the goldfish move in the water?' Maybe you would like to come to the board and show us?

Teacher: Oh, that sort of movement! Movement up and down in the water, not through the water. How shall we make that a little clearer?

Lisa: How does the goldfish move up and down in the water?

Teacher: That's clearer.

                              (Faire and Cosgrove,1988, pp. 20, 21)

CHAPTER 16

Handling children's questions

Notwithstanding the value to children's learning of encouraging their questions, many teachers feel that to do this would only add to their problems. Don't children already ask enough questions and ones which their teachers find difficult to answer? Indeed it is not all that rare for teachers to adopt programmes and classroom strategies which either keep children so busy in routines that they have little time to think and query or else themselves do all the talking and discourage any input from children which is not a response to a teacher's question. So if the ideas suggested in the last chapter for encouraging children's questions are to be taken at all seriously, being able to handle the questions which children raise has a high priority.

Fortunately handling questions is a skill which can readily be developed. It requires thought about the kind of question being asked, about the likely motive for asking it and knowledge of how to turn a question into one which can be a useful starting point for investigation. The word 'handle', rather than 'answer' is used deliberately here. One of the first things to realise – perhaps with some relief – is that *it is often better not to answer children's questions directly* (even if the teacher does know the answer). But it depends on the kind of question which is asked and so we start by identifying important differences.

Questions children ask

Most questions children ask in the context of science activities fall into one of the following five categories which have been chosen because questions in the different categories require different responses.

(i) *Comments expressed as questions*

These are questions which children ask when they are intrigued or excited. The questions don't really need to be answered but there has to be some response which acknowledges the stimulus which gave rise to the question. For example, here is how an infants' teacher handled a question from a six year old when she and a group of children were examining a bird's nest:

> Child: How do they weave it?
> Teacher: They're very clever . . .
> Child: Birds are very clever with their beaks
> Child: Nobody would ever think they were because they're so small
> Teacher: Yes, it's wonderful isn't it? If we turn this right round and let you have a look at this side . . .
> (From *Match and Mismatch: Raising Questions*, 1977)

The child's question was used to maintain the close observation of the nest and a sense of wonder. She might have replied 'Look carefully and see if you can tell how it is done?' but perhaps she judged that this was too early a stage in the exploration for focussing on one aspect. Her response leaves open the possibility of returning to the subject in this vein if the children's interest is still there. Another way of putting this is that she judged the question to be a way of expressing wonder rather than a genuine query. The child might just as easily have said 'Look at how it is woven!'

(ii) *Philosophical questions*

This is another category of questions to which the response has to be of the 'yes, isn't it interesting/intriguing' kind, sharing the wondering behind the question. 'Why do we have birds and all different things like that?' is such a question . Taken at face value the only answer is to say that there is no answer. However, not all children's questions are to be taken at face value; the motive for asking has also to be taken into account (see below). Neither should we read too much into the exact words children use. They often phrase questions as 'why' questions, making them sound philosophical when the answer they are wanting is much more related to 'what makes it happen' rather than 'why does it happen'. When children's question seem philosophical the initial step is to ask them to explain their question. It may well then turn into a question in a different category, but if not it should be treated as an interesting question but not one that we can answer.

(iii) *Requests for simple facts*

These are questions which satisfy the urge to name, to know, to identify. The children looking at the bird's nest asked 'Where did it come from?' 'What kind of stuff is this that it's made of?' 'How long do the eggs take to hatch?' These are questions to which there are simple factual answers which may help the children to give a context to their experience and their ideas about the lives of birds. The teacher may know the answers and if so there is no point in withholding them. In the case of the birds' nest she knew where it had come from and helped the children identify the 'stuff' as hair. But for the length of hatching she did not have the knowledge and the conversation ran on as follows:

Teacher: Well, you've asked me a question that I can't answer – how many days it would take – but there's a way that you could find out, do you know how?
Child: Watch it . . .
Child: A bird watcher . . .
Child: A book
Teacher: Yes, this is something you can look up in a book and when you've found out . . .
Child: (who had rushed to pick up the book by the display of the nest) . . . I've got one here, somewhere.
Child: . . . here, here's a page about them
Teacher: There we are . . .

(ibid)

The group was engrossed in the stages of development of a chick inside an egg for some time. The question was answered and more was learned besides. Had the book not been so readily available the teacher could have suggested that either she or the children could look for the information and report back another day.

Requests for names of things fall into this category, as do definitions which arise in questions such as 'Is coal a kind of rock?' Whilst names can be supplied if they are known, undue attention should not be given to them. Often children simply want to know that things do have a name and, knowing this, they are satisfied. If work requires something to be named and no-one knows the proper name at that moment then children can be invited to make up a name to use. 'Shiny cracked rock', 'long thin stem with umbrella', 'speedy short brown minibeast' will actually be more useful in talking about things observed in the field than their scientific or common names. Later the 'real' names can be gradually substituted.

Some requests for simple facts cannot be answered. Young children often have a view of their teacher as knowing everything and it is necessary to help them to realise that this is not the case. When the children asked 'Where are the birds now, the ones who built the nest?' they were expecting a simple question to have a simple answer. In this case the teacher judged that the kind of answer they wanted was 'They've probably made their home in another shed, but I really don't know for sure' rather than an account of all the possibilities, including migration and whether or not birds tend to stay in the same neighbourhood. A straight 'I don't know' answer helps children to realise the kinds of questions that cannot have answers as well as that their teacher is a human and not a super-human being.

(iv) *Questions requiring complex answers*

Apart from the brief requests for facts, most questions children ask can be answered at a variety of levels of complexity. Take 'Why is the sky blue?' for example. There are many levels of 'explanation' from those based on the scattering of light of different wavelength to those relating to the absence of clouds. Questions such as 'Why is soil brown?' 'Why do some birds build nests in trees and others on the ground?' 'How do aeroplanes stay up in the air?' fall in this category.

They seem the most difficult for teachers to answer but they are in fact the most useful questions for leading to investigations. Their apparent difficulty lies in the fact that many teachers do not know the answers and those who do will realise that children could not understand them. There is no need to be concerned, whichever group you fall into, because the worst thing to do for in either case is to attempt to answer these questions!

It is sometimes more difficult for the teacher who *does* know the scientific explanation to resist the temptation to give it than for the teacher who does not know to refrain from feeling guilty about this. Giving complex answers to children who cannot understand them suggests that science is a subject of incomprehensible facts to memorise. If their questions are repeatedly met by answers which they do not understand the children will cease to ask questions. This would be damaging, for these questions particularly drive their learning.

So what can be done instead of answering them? The best answer is given by Sheila Jelly in the following words:

> The teaching skill involved is the ability to 'turn' the questions. Consider, for example, a situation in which children are exploring the properties of fabrics. They have dropped water on different types and become fascinated by the fact that water stays 'like a little ball' on felt. They tilt the felt, rolling the ball around, and someone asks 'Why is it like a ball?' How might the question be turned by applying the 'doing more to understand' approach? We need to analyse the situation quickly and use what I call a 'variables scan'. The explanation must relate to something 'going on' between the water and the felt surface so causing the ball. That being so, ideas for children's activities will come if we consider ways in which the situation could be varied to better understand the making of the ball. We could explore surfaces, keeping the drop the same, and explore drops, keeping the surface the same. These thoughts can prompt others that bring ideas nearer to what children might do.'
>
> (Sheila Jelly, 1985, p. 55)

The result of the 'variables scan' is to produce a number of possible investigable questions such as 'Which fabrics are good ball-makers?' 'What happens if we use other fluids, or put something into the water?' Exploring questions of these kinds leads to evidence which can be interpreted to test hypotheses concerning what it is about felt that makes it a good ball-maker (and can we use this idea to make it into a poor ball-maker?) and to what extent it is something about water which makes it form balls (and how we can change this). These activities lead towards an explanation of the original question and can be pursued as far as the extent of the children's interest and understanding. It is not difficult to see that there is far greater educational potential in following up the question in this way than in attempting to give an explanation (which probably has to be in terms of a misleading 'skin' round the surface of the drop).

'Turning' questions into investigable ones is an important skill since it enables teachers to treat difficult questions seriously but without providing answers beyond children's understanding. It also indicates to children that they can go a long way to finding answers through their own investigation, thus underlining the implicit messages about the nature of scientific activity and their ability to answer questions by 'asking the objects'.

(v) *Questions which can lead to investigation by children*

Teachers looking for opportunities for children to explore and investigate will find these are the easiest questions to deal with. The main problems are

(a) recognising such questions for what they are
(b) resisting the urge to give the answer because it may seem so evident (to the teacher but not the child)
(c) storing them, when they seem to come at the wrong time.

(a) It is not often that a child expresses a question in an already investigable form; there is usually a degree of 'turning' to do and the 'variables scan' is a useful idea to keep in mind. The example of the snails' shells (p. 121) is a case in point. Here the questions 'Why do snails have four rings on their shell?' was quite easily turned into 'Do snails have the same number of rings on their shells?' A slightly different approach is to turn a question from a 'why' question into a 'what would happen if' question. For instance: 'Why do you need to stretch the skin tight on a drum?' can become 'What will happen if the skin is not tight?'

Not only is this more encouraging for the child than a straight return of the question: 'Well what do you think?' but it directs the child towards finding out more than the answer to the original question – in this case probably the relationship between the pitch of the sound and the tautness of the drum skin.

(b) 'What are these?' (the eyes of the sprouting potatoes)
'Where did these come from?' (winged fruits of sycamore trees)
'How can I stop my tower falling over?' (tower built from rolled newspaper with no diagonal struts).

These are questions which the teacher could readily answer, but in most cases to do so would deprive the children of good opportunities to investigate and learn much more than the simple answer. Certainly, there can be occasions when it is best to give the short answer, but in general the urge to answer is best resisted. Instead it is best to discuss how the answer can be found.

(c) Questions which can be profitably investigated by children will come up at various times, often times which are inconvenient for embarking on investigations. Although they can't be taken up at that moment the questions should be discussed enough to turn them into possible investigations and then, depending on the age of the children, picked up some time later. Some kind of note has to be made and this can usefully be kept publicly, a list of 'things to investigate' on the classroom wall, or just kept privately by the teacher. For younger children the time of delay in taking up the investigations has to be kept short – a matter of days – but the investigations are also short and so can be fitted into a programme more easily. Older children can retain interest over a longer period – a week or two – during which the required time and materials can be built into the planned programme.

Children's motives for questioning

So far we have treated children's questions as if they all stem from curiosity and a desire to understand. These are, of course, powerful motives for questioning and are the basis of the reasons given for encouraging questions (p. 116). But there are other motives which overlay the distinctions between categories proposed above. Children also ask questions to demand attention; they are less interested in the answer than in being the focal point of the class for a few minutes. One little girl made a habit of putting her hand up and then asking a question in such a whispering voice that the whole class had to freeze in order for her to be heard. The teacher did not want to discourage her questions by not allowing her to speak or to be heard, but found the effort disrupted the flow of discussion. Other children use questions to seek gradually more and more clues to what they are expected to do so that they end up with far more help than other children. When such subterfuges work the children use them more frequently and they spread to others. The way to avoid this is not to let them work; to recognise them for what they are and to make it explicit to the child that the teacher realises what is happening.

Recognising that the motive for a question may not be purely a desire to know, the teacher has to modify the ways of handling suggested above. At the same time the situations can be used to reinforce the preference for questions which are investigable, giving praise and attention to such questions whilst expecting children to do more fact-finding for themselves.

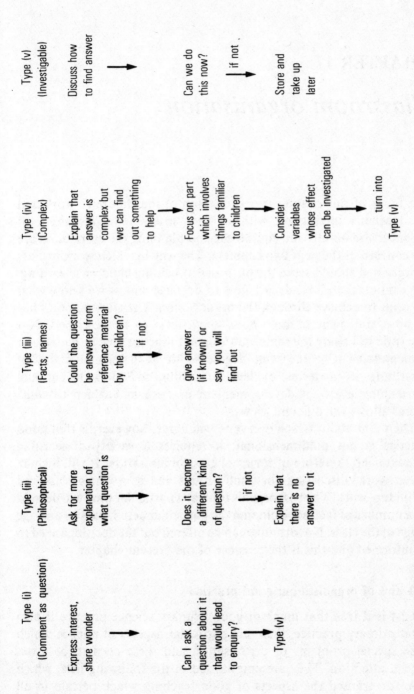

Figure 16.1

Flow diagram for handling questions

Type (i)
(Comment as question)

Express interest, share wonder

→

Can I ask a question about it that would lead to enquiry?

→ Type (v)

Type (ii)
(Philosophical)

Ask for more explanation of what question is

→

Does it become a different kind of question?

if not →

Explain that there is no answer to it

Type (iii)
(Facts, names)

Could the question be answered from reference material by the children?

if not →

give answer (if known) or say you will find out

Type (iv)
(Complex)

Explain that answer is complex but we can find out something to help

→

Focus on part which involves things familiar to children

→

Consider variables whose effect can be investigated

→

turn into Type (v)

Type (v)
(Investigable)

Discuss how to find answer

→

Can we do this now?

if not →

Store and take up later

CHAPTER 17

Classroom organisation

The focus of this chapter is organisation of the children's work and the teacher's interaction with it; the closely related matters of organising materials and equipment and planning the curriculum are the concern of the next two chapters. The way in which work in class is organised should serve the purposes of helping children's learning. We cannot, therefore, decide how to organise unless we know what we want to achieve through the organisation. The starting point has to be a statement of how *good practice* is to be described. Not everyone will share the same view of good practice, nor be capable of implementing it to the same degree so we shall look at a way of describing it in terms of levels (nothing to do with national curriculum levels) of development in practice and which to some extent allows for different views.

The point about which everyone will agree, however, is that good practice is not unidimensional; it requires a variety of learning contexts and therefore of forms of class organisation. It will include group work, discussions in small groups and as a whole class and individual work. The best organisation at a particular time will depend on a number of factors, including the particular activities, size and age range of the class. No formulae can be offered but the decisions need to be informed and this is the purpose of the present chapter.

The aim of organisation: good practice

Whilst is it true that much of good primary science practice is also good primary practice, there are particular aspects of science which pose special problems of organisation and these need to be given special attention. They are emphasised in the following lists, which take for granted the aspects of good teaching which pertain in all subject areas.

The approach to teaching and learning science underpinning this book would lead to a notion of good practice being described in terms of:

- Children's opportunity for practical investigation
- Children's opportunity to develop the whole range of scientific skills
- Children's opportunity to develop the scientific concepts specified in the curriculum
- Children's opportunity to develop scientific attitudes
- Interest of *all* children in the activities
- Relationship of activities to children's everyday life
- Relationship of science activities to other subjects
- Match between the children's initial ideas and skills and the learning opportunities provided by each activity.

These are in rather broad terms but the implications for the classroom organisation, and for programme planning are still evident.

A more detailed list was refined and used by the team evaluating the advisory teacher initiatives for primary science which operated between 1985 and 1988 (under the government Educational Support Grant scheme). The IPSE (Initiatives in Primary Science: an Evaluation) team identified pointers which 'indicate that the science in a school is probably going along the right lines' (IPSE, 1988, p. A1). The statements are arranged in seven levels, the ones with higher numbers being 'possibly only attainable with the most experienced schools and children' (ibid).

I. Children in the classroom

1a. Children undertaking work which is purposeful, and within their experience. By this we mean that the children seem to know what they are doing and why they are doing it.

1b. Children enjoying themselves.

1c. Children showing curiosity, perseverance and self-discipline.

1d. Children able to relate their work to other work, and to everyday experience.

2a. Children discussing their work with each other and with the teacher.

2b. Children working co-operatively with each other, listening to each other, and making genuine decisions together.

2c. Children showing self-criticism and open-mindedness.

3a. Children undertaking investigational work that is genuine, to the extent that they do not know what the 'answer' will be, even if the teacher does.

3b. Children undertaking investigations which arise from their own interests.

4a. Children making predictions about the results that they will obtain.
4b. Children observing systematically, and recording their observations in a systematic and appropriate way.

4c. Children classifying their results and observations, and looking for patterns.

4d. Children drawing conclusions from their observations.

4e. Children using appropriate measuring instruments accurately, and choosing a suitable instrument for themselves.

5a. Children performing and planning experiments in a way that shows understanding of the concept of a 'fair test'.

5b. Children taking responsibility for devising aspects of investigations themselves, leading on with more experienced children to designing and planning with whole of an experiment.

6a. Children handling unexpected results by checking, and repeating measurements. THEN, if results are confirmed, trying to account for them and possibly devising new investigations to confirm their speculations.

7a. Children taking their learning on a stage by looking for further investigations that lead on from their results.

II. The teacher in the classroom

1a. Teacher provides a variety of stimuli and resources for science work.

1b. Teacher ensures that equal provision is made for boys and girls, and encourages equal uptake of opportunities by boys and girls.

2a. Teacher encourages discussion and co-operation between children.

2b. Teacher accepts that a certain amount of noise, mess and movement are inevitable in productive science lessons.

3a. Teacher asks questions which draw out connections and lead on to further investigations.

3b. Teacher uses a variety of appropriate classroom organisational strategies, depending on the task in hand. (e.g. class teaching, groupwork, work as individuals or in pairs, demonstrations, discussions, workcards, etc. etc).

4a. Teacher questions children and discusses their work in a way that helps to relate findings to previous experience in science and in everyday life.

5a. Teacher encourages children to devise their own investigations and places responsibility for aspects of all experiment design on the children.

5b. Teacher questions children in an open-ended way, encouraging them to formulate their own hypotheses, and to devise their own tests for these hypotheses.

5c. Teacher promotes and helps refine the idea of 'fair testing' (however it might be referred to), and encourages children to devise experiments with this in mind.

5d. Teacher is accepting of, and responsive to, children's own ideas, and does not unswervingly pursue some preconceived plan.

6a. Teacher treats unexpected results as a promising source of further investigations, rather than as a 'mistake', and encourages children to do the same.

7a. Teacher expectations of children are matched to their age and ability.

7b. Teacher ensures a balanced content over time between physical science and natural science, and between technological/problem-solving and experimental/explorative approaches.

7c. A balance is maintained over time between the content areas mentioned in the DES/WO policy document (para. 27).
Living things and their interaction with the environment
Materials and their characteristics
Energy and materials
Forces and their effects.

III. The headteacher and teacher within the school

1a. Headteacher is committed to science as an investigational activity, and makes this commitment and her/his expectations clear to the staff.

2a. Headteacher ensures that suitable resources are provided, and teachers have ready access to these resources and use them.

3a. Staff hold curriculum development meetings about (or including) science.

4a. Teacher forecasts or records include work done in science, and show the balance outlined above.

4b. Teacher has access to records which show previous work done by children.

5a. The head facilitates the work of the science co-ordinator, or if such a person does not exist, fulfills the role her/himself.

6a. The head encourages staff to undertake suitable INSET in science, and facilitates this, and there is appropriate uptake of available INSET by the staff.

7a. There is (or is developing) a school policy for science which can aid continuity.

Although the IPSE team did not expect all schools to achieve all of the identified aspects of good practice, they did identify one characteristic which was expected of all, a 'golden rule':

THE GOLDEN RULE – The one absolute characteristic of good

science is that it involves children doing things themselves. This is not to say that there is no place for the demonstration by the teacher, for children consulting books, or even for the occasional straightforward imparting of knowledge by the teacher. We must be careful not to leap to conclusions after a brief visit to a classroom. We need to know what is the overall pattern of activity in a class. BUT, we believe that science must always involve, to a large degree, children actively interacting with the environment, physically handling the materials they are investigating, and not watching someone else.

(IPSE, 1988, p. A1)

If class organisation has to serve the purpose of facilitating the children's and teacher's actions listed here then it obviously has to be varied. But it is probably the Golden Rule that presents the greatest problem to teachers. Children 'doing things themselves . . . not watching someone else' means providing and organising materials and it means organising group work.

Organising group work

Questions have to be addressed about the size and compositions of groups, about how similar or different the groups' work ought to be.

Group size and composition

Group size has to be a compromise between what is desirable and what is possible in terms of the size of the whole class. Groups of four are ideal for the youngest children but juniors can be in groups as large as six if this is necessary, although a smaller number is desirable. The potential for genuine group work, where the work is a combined effort and not a collection of individual efforts carried out by children working in proximity, diminishes the larger the group size but it can also be influenced by the way a task is set up and the encouragement of the teacher.

In relation to the composition of groups there are quite different approaches advocated. For example, a teacher writing in *Primary Science Review* (no 17, 1991, page 8) described her preference for grouping by 'the level of understanding each child has of a particular concept'. In the same publication another teacher wrote 'children of differing abilities would work together easily – they only needed to be able to communicate with each other' .

Clearly very different messages are being conveyed from classroom practice. It may be that the group composition does not actually matter all that much and that other factors (such as the teacher's interventions) are of more significance; or it may be that whatever a teacher feels most comfortable with is the best approach for him or her. One thing that the two teachers quoted would be likely to agree about is that neither had compared in a systematic way the effect of forming groups in different ways.

However, this has been done by a research team and the results are worthy of careful consideration. The initial ideas of the children (about floating and sinking in one case) were assessed and on this basis groups were formed where, in some, initial concepts were different and, in others, they were similar. The groups then undertook the same activities, with the teacher/researcher interacting as would be normal in classroom work, and were compared on the basis of group interactions during activities and of ideas after the activities. The ideas emerging during group activities were assessed from recordings made throughout and the children's ideas were assessed again six weeks after the activities had been completed (and for some topics immediately after the activities as well). Considerable differences were found between groups of different composition:

> Looking at the change from the pre-tests to those post-tests which were administered several weeks after the group task, it is clear that on five of the six comparisons the children from the differing groups progressed more than the children from the similar groups. Moreover, when the differing groups contained children whose level of understanding varied, the more advanced children progressed as much as the less. This being the case we felt our results provide reasonably consistent evidence for the task being more effective with differing groups.
>
> (Christine Howe, *Primary Science Review*, no 13, 1990, p. 27)

What is particularly noteworthy is that the differential progress was *not* apparent during the group work. Neither kind of group seemed to improve on their initial ideas during their group work. In the activities where children were assessed immediately after group work as well as six weeks later, there was no evidence of immediate change in ideas, but this did appear, to a significantly greater extent in the groups with differing ideas, later.

> Thus, there is a strong suggestion that progress took place *after* the group tasks, indicating that interaction when concepts differ is a catalyst for development and not the locus of it.
>
> (ibid)

The researchers followed up the obvious alternative explanations for the differences (for example that further work at school, or experience at home was responsible) but found nothing which could account for them.

The important mesage from this research is that if we judge only from the way the group work goes then there is very little to choose between one basis for grouping and the other. However the exposure to different ideas from their own appears to lay the foundation for greater learning in all children over a longer period of time.

Allocating group activities

The questions here are:
> should all groups be working at the same time?
> should groups be working on the same activities as each other?

The questions are interconnected and so best considered together.

Again practicalities will modify any theoretical ideal. Where specialised equipment is needed and the amount is limited, or requires close teacher supervision, then there is no alternative to having one group at a time using it. What other children will be doing and whether groups will take turns at the same activities remain more open to teachers' decisions.

One school of thought advocates that the amount of science going on at one time is best limited to one or two groups, with others engaged on non-science activities. However, some strong arguments against this are expressed in this quotation from an infants teacher:

> I find that this does not work well. Children are often very stimulated by practical science and discuss their observations excitedly with their co-workers. I certainly do not want to discourage this enthusiasm and the exchange of ideas, but the noise can be distracting for another child involved in a piece of creative writing! Another disadvantage of small group work is that the time for teacher/group interaction is limited, particularly for the important initial discussion and the discussion of results at the end of the session.
>
> (Jane Glover, 1985)

The alternative which this teacher found satisfactory was to have groups working on science all at the same time, with the activities of each being different but all linked to a single theme. The reasons for this being preferred are easy to see, since it solves the problem of distraction and enables class discussions to take place on a shared theme with enough differences between groups to create interest in each others' findings.

This example is a reminder of the multiple characteristics of good practice and the need to balance several requirements. If we go overboard for one (giving special attention to a few groups doing science) we may prevent important opportunities (for discussion, sharing ideas).

Work cards

Work cards can be a useful aid to enabling children to get started at the beginning of a new topic. Activities which invite children to explore materials may be selected from published cards but teachers may prefer to produce their own, particularly if they want to use these activities to gain access to children's ideas. For this purpose a series of questions is enough to start children off. For example, work on soil might begin with 'attention-focusing' questions and 'comparison questions' (see page 110) which could be written down for groups to start on with one or two small samples of different kinds of soil and a hand lens:

> What do you find in the soil?
> What makes the different kinds different?
> In what ways are they the same?

Quite quickly, depending on their previous experience and the ideas which they have, action questions and problem posing questions could be asked:

> What happens if you add water to the soil?
> How would you find out if one is better than another for growing plants?

These questions might be accompanied by oral or written instructions about how to communicate the answers. Varying devices can be used which encourage groups to share ideas, such as:

Get one person to write down all the ideas in the group and then decide which you all agree with and which you don't.

or

Each write down your ideas on a slip of paper and stick them on one piece. Make sure you understand all the ideas even if you don't agree with them.

These kinds of activities provide fruitful starting points for all groups to get to work on, giving the teacher chance to circulate and monitor and decide where to begin to interact and help children to develop their particular ideas. Very soon the activities of the groups will diverge in detail but the teacher should make sure (by 'storing' questions where necessary – see page 128) that they keep to the shared topic.

Organising class and group discussions

Discussion is second in importance only to handling materials as far as value to children's learning in science is concerned. In Chapter 13 the nature and value of discussion among groups and within the whole class has been considered. Here the concern is with setting up the situations which ensure that the potential benefits are realised.

In setting up discussion the (perhaps obvious) point is to ensure the attention of all involved. For a whole class discussion the location of the children is significant in avoiding distractions. When the discussion is a brief one during activities, for the purpose of bringing together observations which have been made, reporting progress, or sharing information which will help everyone (including instructions about safety if unexpected hazards have arisen), it may be advisable to move the children away from the materials they are working on in order to ensure their attention. The discussion will only last a few minutes and it will not be hardship for the children if they are cramped in a small space for this time. It is intended to help them with their work when they return to it, otherwise there is no justification for the interruption.

Apart from these infrequent interruptions, whole class discussion will be at the beginning and end of the session.

The initial discussion is the key to setting up group work which is sufficiently clear and motivating to ensure that children begin work promptly and with enthusiasm. Whether the purpose is for children to continue work already begun or to start out on fresh activity the essential function of the initial discussion is to ensure that children know what they have to do and what role is expected of them. This does not mean that they are necessarily told what to do but where this is not the case, they must know that they are expected to work it out for themselves.

Holding a whole class discussion at the end of a practical session, whether or not the work is completed, should be the normal practice. The reasons for this strong recommendation have been well articulated by Barnes, who points out that learning from group activities:

> may never progress beyond manual skills accompanied by slippery intuitions, unless the learners themselves have an opportunity to go back over such experiences and represent them to themselves. There seems every reason for group practical work in science, for example, normally to be followed by discussion of the implications of what has been done and observed, since without this what has been half understood may soon slip away.
>
> (Barnes, 1976, pp. 30–1)

The teacher should warn the children in good time for them to bring their activity to a stage where it can be put away and to allow five or ten minutes for reviewing and reporting on-going work. At the end of the activities on a particular topic a longer whole class discussion time should be organised and children given time to prepare to report, perhaps with a demonstration, to others.

Group discussions will be part of group work; children should be encouraged to talk freely among themselves. The noise which this inevitably generates is part of the working atmosphere. If the noise level becomes unacceptable it should be possible to spot the reason

- too much excitement about certain activities?
- children waiting for equipment and not 'on task'?
- 'messing about'?

Once diagnosed, appropriate action can be taken – for example, by diluting the excitement by having some children working on different activities, organising equipment for easier access, checking the match of the demand of an activity with the children's readiness to respond.

When the teacher is involved in a group's discussion the purpose may be to monitor progress, to encourage exchange of views, to offer suggestions, to assess. Since it is almost impossible for a teacher to 'hover' without their presence affecting the children, it is best to make clear what is intended. 'I'm not going to interrupt; just carry on' or 'Tell me what you've been doing up to now'. During a teacher-led group discussion, the teacher should show an example of how to listen and give everyone a chance to speak. The group might also be left with the expectation that they should continue to discuss – 'Try that idea, then, and see if you can put together some more suggestions by the time I come back'.

CHAPTER 18

Equipment, computers and other resources

Objects and materials from the environment, and equipment for use in investigating them, are the basic essentials of scientific activity in the primary school. What is fashioned from this basic raw material follows from the role which the teacher plays and the role which the children are given in the learning. The materials have to be there for reasons discussed in the early chapters of this book, hinging on the importance of first-hand experience for children. Three aspects of their provision are considered in the first part of this chapter: selection and storage of equipment and materials in the school; access to those things outside which cannot be brought in and need to be studied in situ; matters relating to safety. The second part looks at the role of computers in science, which is sometimes represented as threatening the first-hand experience of children. However, they have other roles which can make a positive contribution to children's investigations of the world around.

Equipment and materials

The effective use of equipment and materials means that storage and access have to be convenient and efficient from the point of view of time and maintenance. First, however, we shall consider what sorts of things have to be provided and stored.

Selecting equipment and materials

If we have in mind activities such as those with parachutes, with string telephones, with soil and rocks, with plants, with things that dissolve, things that float or sink, the equipment needed is what might be called 'everyday' rather than 'specialised'. A great deal of science work depends on using materials and equipment which will be available in an active primary classroom, such as

> boxes, plastic bottles and other containers, string, scissors, rulers, paper clips, drawing pins, elastic bands, glues, paper, card, plasticine, straws, marbles, pieces of fabric.

Specialised equipment *is* needed, particularly for the activities of older children and for topics relating to electricity, magnetism, reflection, forces and for the measurement of time, mass and temperature. This equipment will include

> torches, mirrors, coloured acetate sheets, glass blocks, triangular prisms, hand lenses, spirit thermometers, night lights, tuning forks, stop clocks and watches, springs, bathroom scales, pulleys, filter paper and funnels, insect cages, aquarium tanks, wood working and gardening tools, magnets, bulbs and bulb holders, wire, crocodile clips.

The list is not exhaustive. It could be very much longer, especially if laboratory equipment were purchased instead of employing 'everyday' things where they can be used. There are good reasons for keeping specialised equipment to the minimum, however, and not just to reduce costs:

> Special equipment that is used only in science and not found in other parts of children's school or everyday experience can isolate science from the 'real' world around children. If a special set of instruments has to be used for weather observations, for example, the impression may be given that useful measurements depend on that set. It can come between the things being measured and the child. A better understanding of what is being measured may come from a home-made rain gauge, a wind sock made from a stocking and an anemometer made with yoghurt containers rather than from more sophisticated equipment.
>
> (Harlen, 1985, p. 221)

A third type of material that is needed is consumable; the flour and soap powder used in dissolving activities, the wood which is hammered, the wire used up in the home-made sonometer, the batteries, the aluminium foil, the fruits which are cut open, the seeds which are sown, etc. Shortage of these things causes the greatest frustration because they are what many of the children's activities are all about – the stuff of exploring materials from their surroundings. Some of these things can be obtained free, but by no means all. Furthermore, they cannot all be bought at one time of the year and stored. Schools need to set aside sums for these purchases and to establish, generally through the science coordinator, a means of access to and control of, the funds.

Storing equipment and materials

Access is the key word in deciding a system of storage for equipment and materials. There are various possibilities and the advantages and disadvantages of each in a particular case will depend on the size, physical layout and curriculum planning of the school. We can do no more here than point out options.

A decision has to be made about central storage versus distribution of the equipment among classes. Apart from the physical availability of a central store a major consideration is having someone to look after it. There are obvious advantages in sharing expensive items which are only infrequently used but some of these advantages are lost if the equipment is not kept in good order. Clearly the science coordinator has to be willing and able (in the sense of having the time to devote) to organise a central store and to check that items are not 'lost' by being put back in the wrong place or in an unsatisfactory condition.

Another decision is whether pupils should have access to the equipment as well as teachers. The problems of maintaining an orderly central store can be exacerbated by too many having access, yet the teacher will want pupils to help in the collection and return of equipment. The suggestion of appointing a few children as 'monitors' or 'storekeepers' may be a solution. If the store is within each class the same considerations apply. If children are to have access then the labels used to classify the equipment should be ones that they will relate to and understand.

Whether or not there is a central store, within a class the equipment for a certain session needs to be accessible to the children. The demands of providing group activities for all the children at once are, of course, considerable and require preplanning and preparation. Without limiting what the children can do, which should emerge from the initial discussion and their own ideas, the materials and equipment needed can be anticipated and a suitable selection made available. From then the teacher should be able to depend on the help of the children to take responsibility for choosing, collecting and later returning the equipment they use. Building up a system for this is important for the children's ability to take a part in facilitating their own learning as well as for the teacher's sanity. It involves making sure that children know what is available, where, and how to look after it and keep it tidy. There are considerable dividends for the initial investment of time when children are, perhaps, involved in drawing up lists of what equipment there is and creating rules for using the store.

A third major decision point, which applies where a school or class organises science within topics, is whether equipment should be boxed by topic or stored as separate kinds of items. The topic box is a great convenience, but can tie up equipment which could be used for work outside the topic. This can lead to 'plundering' from the box with the chance of the item not being there when that topic is being used. The effort put into developing topic boxes is also a disincentive to changing topics, when perhaps they have outlived their freshness. The device of temporary topic boxes is a compromise. The box exists for as long as the topic is being used and is dismantled when moving on to another topic.

Access to materials outside the classroom

When the children have to be taken to materials for study rather than the material brought to the children, the first step is for the teacher to make a visit well in advance of taking the children. Whether the place to be visited is a wood, a piece of coastline, a small factory or workshop, a church, an airport or a row of houses or shops there will be an immense wealth of possible things to notice, to do and to follow up. It is essential for the teacher to consider it all and plan out beforehand answers to the following questions:

- What do I want the children to notice?
- What do I want them to investigate (on the spot and back in the classroom)?
- Which ideas do I want to help them develop?

Based on the answers to these questions, more detailed planning needs to include:

- Questions to ask the children to stimulate their observation, their investigation, their questioning (as suggested in Chapter 15)
- Preparation in terms of skills they will need (such as the use of a hand lens or pegging out a minifield)
- Equipment they need to take with them.

The school will have procedures and regulations relating to out of school visits. In additions to following these rules, there must be time before the visit for the teacher to work with the children to

- set the scene
- collect all the necessary equipment and ensure the children know how to use it
- explain, and insist upon compliance with, safety measures, parts of the countryside code and considerate behaviour
- introduce sources of information for use in the follow up work after the visit.

Safety

Fortunately science at the primary level is not, in practice, a source of danger to children; very few accidents related to science activities have occurred. However, now that virtually every teacher is involved in science and technology activities, there is a need for existing precautions to be spread more widely and to become established in routine practice. With this in mind the Association for Science Education has recently produced a new edition of its publication *Be Safe* (1990) which brings together the best advice which primary teachers can find on the subject.

Safety is not only a matter for teachers to consider in obtaining and using equipment; these considerations should be shared with children. The national curriculum now requires explicit development by children of ideas relating to road use, mains electricity, health hazards or smoking and the abuse of drugs and solvents. Children need to come to an understanding of the dangers these present so that self-discipline replaces obedience to rules. Rules and obedience to them is necessary where safety matters are concerned but the sooner compliance becomes voluntary the sooner the temptation to break them is eliminated.

Developing a willingness in children to use their understanding and to take sensible safety precautions requires the teacher to take the same role as in the development of other attitudes. The suggestions in Chapter 12 apply, with particular emphasis on setting an example. The prime importance of safety should not operate to curtail children's investigations but to ensure that the necessary precautions are taken and that children gradually come to understand the reasons for them.

Using computers

The use of computers in primary science has not developed as quickly as was perhaps anticipated in the early 1980s and many of the issues which were debated then have not been resolved through practice. Nevertheless there has been sustained work by a number of practitioners which has provided some examples of what can be done and enabled advantages and disadvantages to be better appreciated.

The main uses for (micro)computers are for simulations, data handling, information technology and word processing. This leaves out purposes of control technology, although there is a strong overlap of technology and science under the heading of information technology.

Simulations are most commonly used at the secondary level and many people are sceptical about their use in the primary school, fearing that they will replace first-hand exploration and investigation. On the other hand there is a greater welcome for them in providing information which is not easily available to children in any form, for example, about the movements of stars and constellations. In the introduction to a book which discusses the pros and cons of simulations helpfully and at much greater length than is possible here, Jan Stewart states that

> If used alongside rather than instead of first-hand experience they may have a place. Where they only provide more tedious experiences with fewer decision-making opportunities than real life counterparts they should be rejected. Certainly, at their best, simulations do provide opportunities for observation, the development of concepts, logical thought, the posing of questions and the selection of answers plus opportunities to note patterns and relationships. But at their worst they can become the means of a mechanical convergence towards the discovery of the right answers; a mere game in which children develop 'winning strategies' but fail to understand the essence of the event simulated.
>
> (Jan Stewart, 1985, p. 10)

Used creatively by the teacher, many simulations which could be mere games, can have a greater educational value. Children can make up challenges for each other, for example, or be obliged to predict and justify their prediction before trying it out on the computer. This points to the unavoidable fact that the teacher remains the main medium of education; putting children in front of a screen is no substitute.

Data handling chiefly involves the use of data-base programs. These require children to sort out and organise data so that they may then pick out patterns and make predictions. Various programs allow the user to define their own field relating to the variables of an experiment. Many are suitable for children from the age of 7 or 8.

For example, in an investigation of parachutes (reported by John Meadows, 1988) the fields included the independent variables of the canopy material, the canopy shape, canopy area, the number of strings, the length of strings, the material of the strings, the mass of the load added, and the dependent variables relating to the way the parachute descended, the stability, direction and rate of fall. Once data had been entered about all the models used in the class the data base was used to answer questions such as:

> Is there a discernible pattern between the time for descent and the canopy shape, length of string and load mass?
> Is there a relationship between load mass and stability?
> What factors seemed to be related to stability?

This example seems to show that the main benefit from the considerable work of setting up this data base would be the appreciation of what a data base is and how it can be used. Once established as a tool, however, and used in other contexts, it can help children to test ideas about associations between variables and so further their development of understanding of the things being investigated.

Information technology in this context means using the computer to solve problems of obtaining data as opposed to processing it. Some data is difficult to collect (because things happen too quickly for human reactions, for example) and some requires tedious repetition of observations. An example of the later is the observation of animal behaviour under different conditions. Clough (1985) describes the example of a temporary home for a mouse being constructed so that the mouse's movements could be followed. The device involved pressure pads at the four corners, forming switches which sent signals in various combinations to a computer programmed to turn these into a plot of the position of the mouse. Data could then be collected around the clock. Other methods of detecting movement, fast and slow, use interrupted light beams and light sensitive resistors. Clearly these activities are for top juniors and the teacher who is an IT enthusiast.

Word processing is becoming familiar in work across the curriculum, where the opportunities it affords children to edit and redraft their writing provide a more satisfying product without the drudgery of rewriting. The particular advantages for science seem to arise when children begin to use word processing as routine. Their report on their work then becomes more than a piece of writing at the end; it is more integrated in the whole of an investigation. A primary head teacher describes the change as follows:

> when the writing up stage is seen as an integral part of the scientific process it can be used to develop an awareness in children of the importance of attention to detail and to promote a more thorough understanding of what they are doing . . . Before word processing, relatively little was done in terms of planning and improving scientific writing in spite of regular encouragement by teachers to do so. The need to make notes and extract information tended to be overlooked in the excitement of the investigation and the eagerness to get on.
>
> (David Clough, PSR No 5, 1987, p. 5)

CHAPTER 19

Curriculum Planning

Curriculum planning in the school takes placed in three stages – the whole school level, the class level and the individual pupil level. The last of these is a matter of matching activities to individual children and then of keeping track of their activities, things which are discussed in Chapters 9 and 24 respectively. Here the focus is upon the first two stages, where the aim of the planning is to find a way for all the children in the school to have the opportunity to encounter all the relevant parts of the programme of study and through it to achieve learning at the levels of which they are capable.

Before planning at school or class levels can proceed, there are some issues which have to be faced and which depend, not on the national curriculum, but on the educational values, preferences, strengths and weaknesses of the staff of the school. These include the pros and cons of different forms of topic work, the extent of spiralling in the curriculum (revisiting attainment targets), timetabling and how the achievement of process skills is to be accommodated. We shall discuss these issues briefly before giving some examples of approaches to planning at the two levels.

Issues relating to provision for science

Topic work means that different aspects of work are linked together so that they reinforce one another and are studied within a context which has meaning for children in relation to their experience. The different forms of topic work arise from the breadth and variety of what is included. Cross-curricular topics link two or more subject areas with equal emphasis on the components. Science-based topics have a focus on science and, while other areas of the curriculum are inevitably involved, their role is incidental rather than planned.

The advantages of topic work are well expressed by two experienced science advisers who first explain it as:

> an interdisciplinary theme taken by the class for a number of weeks. Groups of children work individually and cooperatively at different aspects guided by the teacher. There are opportunities for the development of class and small group teaching, investigations, visits, and displays. Such an approach encourages children to follow their own interests and design their own investigations. It also allows them to plan their work and share their ideas. The teacher is able to provide a variety of experiences at an appropriate level and to have dialogue with the children, helping where necessary. This way of working avoids severe problems of timing and shortages of equipment, and avoids the artificiality of subject 'boundaries'. Teachers working in this way find it relatively easy to introduce science based on skills and problem solving.
>
> (Barry Davis and John Robards, PSR, No. 9, 1989, p.8)

This is fine as long as teachers *are* able to provide all these experiences and appropriate support for children. But what if they can't? The disadvantages of topic work for learning science hinge on the considerable demands of working in this way and the expertise required of the teacher. HMI pointed out in 1989 that 'over-ambitious topic work, in which too many elements and subjects are attempted at the same time, trivialises the children's learning in science and fails to establish the aimed-for connections between science and the various constituent subjects of the topic' (DES, 1989, p.18–19). Whilst acknowledging the benefits of linking work in different subjects and that science can be linked with virtually every other subject, they caution that:

> the least effective work is often associated with topics where far too much is attempted and – as a consequence – too little is achieved in depth of knowledge, understanding and the acquisition of skills in the constituent subjects. More often than not standards were depressed in these circumstances because the teacher could not cope with the complexities of managing the work on such a broad front.
>
> (DES, 1989, p.20)

Whether or not parts of the programme of study for science are encountered through topic work, there has to be decision about how often each is 'visited' in the planned work of a child as (s)he passes through the school. At one extreme would be to cover the whole range of the programme of study and attainment targets every year, with progressively more complex ideas being built on those before. At the other extreme would be to work though the programme of study once during the infant or junior years. An annual cycle has disadvantages of having to cover too much ground each year, which would obstruct exploratory and in depth work. On the other hand a four year cycle is far too long; theoretically, some concepts might have been encountered only when a child was aged seven or eight and others only at age eleven. A two year cycle is a good compromise and this is illustrated and elaborated in an example given later.

The timetabling of science activities is an issue confounded with that of topic work. Where topic work is the predominant way of working, the timetable usually allows this to take place for extended periods of time, interrupted only by the essential scheduling of activities where space, staff or equipment have to be shared among all the classes. Thus there is time for children to carry out investigations which would not fit into small time slots. If the school organises curriculum provision on a subject basis, however, with the time for each area designated, there is a greater likelihood of time being chopped up into portions which restrict opportunities for children to try things out, discuss them, try other ideas whilst things are fresh in their minds and so derive maximum learning from their activities. True, there can be the 'afternoon for science' to allow this extended time, but then the problem is of continuity from one week to another.

Although the time to be spent on science is not prescribed, the general understanding is that it should be at least one tenth of lesson time. This is more easily computed when 'science lessons' are stipulated but this should not be a reason for separately teaching science or indeed other subjects. Teachers do, however, have to consider time in their planning, since it is very easy to 'lose' the science in a topic approach and so spend far less time than is needed for meeting the requirements of the national curriculum.

The position of planning for process skills is different from that of planning for the knowledge and understanding outcomes of learning. These process skills have to be used in all activities and so do not appear in planning topics or other curriculum schemes at the school level. It is important that they are not forgotten, however, and so a school's policy for science should include explicit statements about how they are to be included in planning, predicting and so on. If other activities include the process skills, there is no harm in this approach, but the danger is when there is an assumption that process skills have been 'covered' by these special exercises and then the rest

Year	AT1 Scientific investigations	AT2 Life and living processes	AT3 Earth and environment	AT4 Materials and their behaviour	AT5 Energy and its effects
6	✔	✔		✔	
5	✔		✔		✔
4	✔	✔		✔	
3	✔		✔		✔
2	✔			✔	✔
1	✔	✔	✔		

Figure 19.1

of the curriculum can be delivered through formal 'transmission'. It is important to recall that process skills are used in activities in order for children to develop their ideas and learn with understanding (as was explained at length in Chapter 2).

Planning at the school level

Different approaches are taken to deciding how the programmes of study are distributed among the classes in a school. They all have advantages and disadvantages which can most easily be seen by regarding them as variants of two main approaches.

Approach 1

In this the parts of the programme of study and associated attainment targets are distributed to each year and teachers may then plan to meet these in whatever way they choose – through science activities related to science themes, science-based topics or cross-curricular topics. In national curriculum terms, based on a two year cycle, the distribution might look like Figure 19.1.

Of course it isn't necessary for the whole of an attainment target to be the unit for allocation and these can be subdivided and parts assigned to different years. The advantage of this approach is that teachers have a considerable amount of autonomy in relation to how they arrange their work, but this brings the associated disadvantage of allowing possible duplication or overlap in topics. This can, of

Year 3 Topics	AT2	AT3	AT4	AT5
Homes			/	○
Whatever the weather		/		●
Sound and music	/	/	●	
Plants and growth	●			/
Shape and pattern	●		●	
Switch on the electricity		/	●	
Year 4 Topics				
Ourselves	●	/	/	/
Communications	/		●	
Reflections			●	/
The iron man		●	/	○
Journey into Space			/	●
Water	/	○	●	/
Year 5 Topics				
Materials		●	/	○
Changing life	●	○		○
Using electricity			●	/
Round about Tottenham	○	●	/	○
Toys and games		○	●	
Year 6 Topics				
Beneath our feet	○			●
Light and shadows			●	/
On the move		○	●	
The Earth in space				●
Keeping healthy	●	○	/	

/ = covered briefly   ○ = covered moderately   ● = covered in depth
(Based on article by Andy Waterman, PSR, 15, 1990, p 18)

Figure 19.2

course, be avoided by teachers putting together their plans and by keeping records of the activities of classes and individual children.

Approach 2

This is designed to avoid duplication of topics by agreeing these for the whole school and mapping attainment targets to them so that they are all covered. In recognition that topic work will cover more than one area of science, the example here elaborated the approach by identifying attainment targets which would be covered briefly, in some detail and in depth. Topics were decided, by the whole staff working together, for each half term in every class on the junior school. Figure 19.2 indicates the scheme which emerged.

The staff of this school invested a great deal of time in their plan and indeed also put school capitation into developing 'topic boxes' which would include the special equipment needed for each topic plus other useful materials, pupils' books and so on. It seems likely, then, that this plan is intended to operate year after year; indeed it would be difficult to interrupt its operation without rethinking the whole scheme and taking into account what the children had already done. Schools would not have time for this every year and so it would become the routine that, for example, each first year class would begin with 'homes' and then go onto the 'weather' after half-term. The obvious disadvantage is the possibility of staleness which might invade the work. On the other hand all topics on 'homes' are not the same and teachers could be encouraged to introduce variety into the way they tackle the work.

(Sue Davis, *PSR*, No 11, 1989, p. 19)

Figure 19.3

Planning at the class level

The aim at this stage of planning is to work out activities within topics, if that is the intended approach or otherwise for science themes, which enable the appropriate part of the overall school plan to be implemented. There are various stages to go through which can be called

> brainstorming
> analysing
> focussing
> developing.

Brainstorming is the creative step when all kinds of ideas for work on a topic or theme are put down without too much regard for practicality and detail. These considerations will come in later. The work may lack sparkle if ideas are restrained too early. Ideas will come from published materials, from other teachers – there are plenty available on almost any topic or theme. Brainstorming results in a wealth of ideas readily expressed in a flow diagram, such as in Figure 19.3.

The next step of *analysing* means deciding what potential for learning there is in each of the activities. This might be done by attaching attainment target or strand labels to each one or developing a matrix so that each can be related in more detail to parts of the programme of study and to the attainment targets. In this process, the activities have to be identified in more detail and some may be eliminated as a result of thinking through what the children would, or would not, achieve from them.

The process of *focussing* is then already beginning, when decisions are made about which activities, of all the possible ones, will be developed. Many considerations enter at this point: practicality (what can we learn from a visit to the park, given that we can only visit once?), coverage of the programme of study (taking out those which are already well covered and leaving those for which there are fewer opportunities), potential for achieving attainment targets (those which are planned for this part of the year and which include statements at the appropriate levels).

Then comes *developing*, meaning deciding the starting points for children's activities. These activities, at the beginning of a topic, should be ones which give children chance to explore objects and materials relating to the topic so that the teacher can find out their

Class .... Topic .... Date ....

Activities	Parts of Programme to Study relating to:															
	AT2				AT3				AT4				AT5			
level	2	3	4	5	2	3	4	5	2	3	4	5	2	3	4	5
....		□	□							□	□			□	□	
....							□							□	□	□
....		□	□											□	□	
....									□	□	□	□				
	□	□	□	□												
							□							□	□	□
....			□				□	□	□	□	□					

Figure 19.4

ideas. Subsequent activities which help children to develop their ideas can be planned in outline but the details have to be decided in response to the children's ideas.

Keeping track of plans and practice

In the real world there will always be a difference between what is planned and what happens in practice. It is, therefore, necessary to keep records of what is implemented. It is useful if this is done on the planning sheet emerging from the focussing step, as described above, since then the gaps not actually covered can be readily seen. Figure 19.4 is a possible format.

Here the open boxes refer to what was planned. What actually happened in practice is indicated by some mark in the boxes. Where there is a difference the teacher can consider whether action needs to be taken. It may be that a significant part of the topic was missed for some reason and provision has to be made for it, even if this means some activities which are more teacher-directed than usual. More likely, however, is that the missed opportunities for developing ideas can be built into the planning of a future topic. Generally, too, there is some planned overlap between activities in various topics, intended so that children can meet scientific ideas in different contexts, and so what is missed may be made up by taking advantage of this. Keeping and reviewing the records is an important basis for such decisions.

CHAPTER 20

About assessment

This chapter forms an introduction to concepts and issues about assessment in general and a basis for the following three chapters which deal with matters more specific to assessment in science. The main focus in the following chapters is informal assessment which is part of teaching, rather than testing. But first we put both informal assessment and formal testing, and other varieties in between these, into the context of a discussion of the meaning(s) of assessment, its purposes, range of methods and bases for making judgements. Finally there is a brief discussion of national testing and assessment.

Meanings

It is generally agreed that assessment in the context of children's achievements in school is a process of making judgements about the extent of these achievements. The judgements are reached on the basis of information which has been gathered about performance and which is compared with some kind of expectation. The various ways in which information is collected and the various bases for judging it create the variety of different kinds of assessment. These include the gathering of information through carefully devised standardised tests, under controlled conditions and, in contrast, ongoing assessment, carried out almost imperceptibly during normal interchange between teacher and pupils.

The major distinction within assessment methods is between tests (and examinations) and other forms of assessment. Indeed some use of the term 'assessment' excludes tests and means only various forms of informal assessment usually devised by, and always conducted by, the teacher. Tests are specially devised activities designed to assess knowledge and/or skills by giving different pupils precisely the same task under conditions defined by those who devised and trialled the test. However the distinction between tests and non-test assessment is not always all that clear. Some 'tests' can be absorbed into classroom work and look very much like normal classroom work as far as the children are concerned and so they cannot always be regarded as 'formal'. To be more useful, the distinction should go beyond methods to include purposes. The main purpose of tests is to check what children have achieved, although in some cases they also serve a purpose of feedback to help learning.

Purposes

Asking teachers why they assess pupils reveals a wide range of reasons and shows what an important part assessment plays in teaching and learning. For example the following list was quoted as typical:

- To gather information about a child's progress through a scheme of work
- To identify when a child is having problems
- To devise future work schemes
- To see if teaching is effective
- To satisfy the criteria of school/LEA documents
- To collect information for support services
- For parents' evenings
- To avoid repetition of work
- To ensure continuity of education
- To ensure progression of learning

(David Palmer, *Primary Science Review* No 7 p. 26)

These and similar purposes can be organised under headings such as the four used in the Report of the Task Group on Assessment and Testing (TGAT) DES (1988c): formative, diagnostic, summative and evaluative. The first two of these can usefully be combined, as we shall see if we look at the meanings of these categories.

(*Formative* assessment is aimed at helping the teaching and learning process; information gathered regularly is used for making decisions in on-going work. It assists teachers in adjusting the challenges given to children to match their existing ideas and skills, to help rather than to grade children. It is usually informal in that the child is not aware that it is taking place.)

The term *'diagnostic'* is sometimes used as if this were different from formative assessment. Diagnostic assessment has a more specific focus, being concerned with examining in depth a particular area of performance. But this is only a slight variant of formative assessment and can be considered part of it.

Summative assessment, as the name suggests, means a summary judgement or a summing up of where a child has reached at a certain time. Quite often the information is obtained by a test (or examination) at the end of a term, year, or a certain section of work. But it is also possible to give a summative assessment as a result of reviewing records of on-going assessment, as teachers frequently do in reporting to parents, either orally or in writing, at the end of a year.

Summarising information which has been gathered and used for formative assessment means that the summary has the breadth and detail of the record created over time and the advantage of not requiring the teacher to gather any more information. It can be done by checking through the records with the help of a record sheet so that there is some uniformity in what is included in the summative report for each child. The disadvantage of this approach is that some of the information may be out of date, if parts of work visited earlier in the year have not been revisited giving the teacher opportunity to update information.

The alternative approach, providing summative assessment information by giving a test to check up on the current performance of children near to the time of reporting has the advantage of being up to date, but the disadvantage of a necessarily restricted range of information. The notion of providing summative information through a mixture, partly based on rich but possibly dated formative records and partly on a final check-up test, is attractive provided that there is a way of dealing with any conflict between information of one kind or the other. (Later we shall mention this conflict in relation to National Assessment in England and Wales).

Evaluative assessment is a term used in the context of the National Assessment in England and Wales to mean providing information about the achievement of *groups* of pupils for reporting on the performance of the school, or larger unit such as the education authority or the national system.

The information in this case is not used for any decisions about the pupil and so pupils can be anonymous. A survey approach, as used in the Assessment of Performance Unit (APU) monitoring of performance in England, Wales and Northern Ireland (now terminated), and the Assessment of Achievement Programme (AAP) in Scotland (still continuing), serves the purpose of providing information at the regional and national levels. As the APU surveys showed, a detailed profile of performance of the national or regional population can be created by assessing a sample of pupils and without requiring the whole sample to take the same test items.

As we shall discuss on page 165, the National Assessment proposals include using information collected about individual pupils for summative purposes to provide evaluative information about the performance of groups – classes or schools, or authorities or the whole population of a year group. Using information gathered for one purpose to serve another in this way introduces tensions into assessment of pupils and we shall return to this later. For the moment it is necessary to point out that information about pupils' performance on its own does not allow a fair assessment of the effectiveness of teachers and schools. Other information, about the socio-economic background of the children must be provided and taken into account, since there is massive research evidence that children of socially disadvantaged backgrounds tend to have lower achievement than those of more advantaged backgrounds (Gipps, 1990, p. 40).

Methods

Ways of collecting information about children's achievement of ideas, skills and attitudes can be categorised in terms of what the children are doing when the information is collected and how the information is collected. The children may be engaged in

— normal work (including both written and practical work)
— special practical tasks (including tests)
— special written tasks (including tests)
— self-assessment
and the teacher may be
— observing, but not interacting with, the children (including watching and listening)
— interacting with children (as well as watching and listening)
— using a check-list
— marking tests
— reading or marking class work
— gathering general impressions.

Common combinations of items from these lists describe identifiable 'methods' of assessment such as tests, continuous assessment and ratings, but there is clearly a range of other possibilities. Some methods are more suited than others to collecting information about achievement in particular subject areas and in Chapters 22 and 23 we shall focus on methods particularly appropriate for performance in science.

Basis of judgement

This refers to the reference point or system used in judging information. It may be illustrated in terms of an example. Suppose that a teacher wants to assess a child's ability in knocking nails into wood. The teacher may have some expectation of the level of performance (knocking the nail in straight, using the hammer correctly, taking necessary safety precautions) and judge the child's performance in relation to these. The judgement is made in terms of the extent to which the child's performance meets the criteria; that is, it is *criterion-referenced*. Alternatively the teacher may judge in terms of how the child performs at knocking in nails compared with other children of the same age and stage. If this is the case there will be a norm or average performance known for the age/stage group and any child can be described in relation to this as average, above average or below average, or more precisely identified if some quantitative measure has been obtained. (The result could be expressed as a 'knocking nails age' or a 'hammer manipulation' quotient!) The judgement arrived at in this way is called a *norm-referenced* assessment. A third possibility is that the teacher compares the child's present performance with what the same child could do on a previous occasion – in which case the assessment is *child-referenced*, or *ipsative*.

It is important to recognise these different bases for judgements in assessment and apply them appropriately. Each has its value in the right context but each has drawbacks outside these contexts. Child-referenced assessment is appropriate for formative assessment, for providing encouraging feedback to pupils, particularly slower ones who, if compared with criteria or with others' performance would always be seeming to fail, but can recognise progress in terms of their own previous performance. But it must be realised that it leads to one child being praised for work which, from another child, might be received with less approval. This is no problem as long as no comparisons are made between children, but where comparisons are being made, or performance in terms of external standards has to be reported, then one of the other bases for judgements must be used.

Interpreting results of assessment

Results of assessment have to be interpreted in the knowledge of the kind of information gathered and the basis of the judgement made. There is always some implied generalisation of the result, for we assess only a sample of behaviour and act on it as if it applied to more than this sample. Indeed unless we are assessing simple recall of facts, we have the expectation that the result will tell us more than just about the particular performance assessed. However, it is important not to generalise beyond what is justified by the information. But how is this limit to be identified? The problem can be put in this way: if a child has shown evidence, either in a test situation or in regular work, of using patterns in findings to make predictions, to what extent does this mean that (s)he is able to do this on all other occasions where it is possible?

It is widely recognised that the context and the subject matter (which has an influence on motivation) of tasks affect performance. Some children anticipate failure in certain activities because of their self-image (see Chapter 5) and so make failure more likely. The poor performance of girls in some activities involving ideas of physics can be explained to an extent by the context and topics having a greater identity with boys' interests and triggering the reaction in girls that 'I can't do this' before even trying.

The APU surveys in science used a large number of items to assess achievement in each of several categories of performance in science and so are a good source of information about how performance varies when attempts are made to assess the same skill or concept application in different contexts. Wide variations in average performance were found which were difficult to explain, although it was possible to speculate about the causes (DES, 1985). Sometimes children performed better when the use of the scientific skill or idea was set in the context of everyday activity, whilst this kind of setting was associated with a depression of the level of performance in other cases. This is not a source of variation found only in science. Measured performance in mathematics depends strongly on the context in which the mathematics has to be used. Boys regularly calculate batting averages in cricket but experience difficulty with the same calculations when decontextualised. Research in mathematics has shown a factor of three in the difficulty of different questions devised to assess the same mathematical skill (Wilson and Rees, 1990).

So we know from this evidence that context does influence performance but to an unknown extent. Assessment results must therefore be interpreted cautiously, as guides to what children can do but not as indicating any kind of certainty about it. No assessment can be used predictively with certainty; it is best treated as a hypothesis, a tentative finding, to be modified by further evidence of the child's performance.

The kind of evidence which is the basis of the assessment is the best guide to interpreting results. Even in such apparently generalisable assessment as is derived from standardised reading tests it is always necessary to specify the title of the test in quoting reading 'ages' since different tests differ in the emphasis placed on the various aspects of reading.

Cautious interpretation should avoid labelling children, which results from over-generalisation of assessment results. When an assessment is taken to describe the whole child and not just a certain performance this can affect teachers', parents' and the child's views of what the child is able to do, often needlessly limiting expectations.

National assessment and national testing

The National Assessment programmes of the UK have the following characteristics:

(a) *They are criterion-referenced.* The criteria are the statements of attainment in the national curriculum of England and Wales and Northern Ireland (these are called attainment targets in the Scottish 5 to 14 Development Programme).

(b) *They are curriculum-related.* This follows from the above, but it is worth pointing out the difference from assessment which are not curriculum-related, such as VRQs, reading ages and other standardised test scores. The essential distinction is that a curriculum-related assessment indicates which parts of the curriculum have been mastered and what has yet to be achieved.

(c) *The results are expressed in terms of 'levels of achievement'* (10 levels in the national curriculum and five in the Scottish 5 to 14

Programme. Using these level labels is a short-hand for the criteria at that level, but there is an obvious danger in using numerical labels as if they were numerical values. This is avoided in the Scottish curriculum where the levels are labelled A to E.

(d) *Include Teachers' Assessments and Standard Tests*. Teachers' assessment is carried out throughout the year and uses information mostly gathered from regular class activities rather than specially devised situations. It collects information from the whole range of contexts of children's activities. It is an integral part of teaching and learning and so should not take up additional time. Standard Tests (SATs as they are known in England and Wales) take place at certain times only (at the end of Key Stages in England and Wales and during the years P4 and P7 in Scotland). They are externally devised tasks designed to give information about performance in relation to a sample of attainment targets, but administered and marked by the teacher. They inevitably take time from regular work and are confined to performance in a few contexts.

(e) *The same information is used for different purposes*. This applies in England and Wales only, where the TGAT report proposed that the same information should be used for formative, summative and evaluative purposes, by aggregation. Thus summative results for individual pupils can be formed by aggregating formative results, to give profile component or whole subject levels. Then summative results for individuals can be combined to give results for classes, or for a whole school or even across schools to give levels for an age group.

(f) *There are procedures for moderation*. In England and Wales moderation has three components. The first takes the forms of 'agreement trials' where groups of teachers consider the same examples of children's work and share their judgements. The aim is to improve the comparability between teachers' judgements both during teachers' assessment and marking SATs. The second and third components apply to the SATs and involve moderators who visit schools during SATs and may assist in procedures for resolving conflicts between the result of the teachers' assessment and the SATs.

It is evident from this account that there are some significant differences between procedures in England and Wales and in Scotland (the procedures for Northern Ireland have not been finalised at the time of writing). Two aspects of difference merit further comment.

1. *Combining teachers' assessments and SAT results.*

In England and Wales, a single result for each attainment target is reached at the end of each Key Stage from the teachers' assessment and the SAT. Where there is conflict between the two results 'in general, the SAT assessment is to be used' (SEAC, 1991). This ruling assumes that the two assessments really should lead to the same result and that the teachers' assessment is less reliable than the SAT where they are not the same. However, since the nature of the assessments are different, not least in one being spread across a number of occasions familiar to the children and the other depending on a single, carefully prepared, task set in one context, then it should not be assumed that they will be the same. Indeed the early pilot trials showed that 'In science, the teacher's rating differed from the test's for nearly 70% of the areas tested. Teachers and tests were at odds for nearly 60% of the areas in maths, and for 43% in English. In most cases, though not in adding and subtracting, children did better in the tests' (Independent on Sunday, April 14th 1991). The question of which is 'right' is not answerable and it would be better to report the two separately (as proposed in the Scottish procedures). The decision to overrule the teachers' assessment has a damaging effect on teachers' confidence and helps to give the impression that the only worthwhile assessment is formal and separate from teaching.

2. *Using information for more than one purpose.*

If information is to be used for formative as well as evaluative and summative purposes, it has to be in sufficient detail to serve the purpose of formative assessment. This means that sensible ways of aggregating results have to be found. This is not easy and various kinds of anomalies can be found in the rules proposed. It is not this aspect of using the results which is the main worry, however, but the principle of using pupil performance results alone for evaluating the work of teachers and institutions. The unfairness of comparing schools in very different social areas on the basis of pupil performance is compounded by the effect that this might have on enrolment and, through that, on the financial support of the school. The effect on teachers is to make assessment seem threatening rather than supporting to their work. This is heartily to be regretted since, as we hope to show in the next chapters, assessment can and should be of benefit to both children and teachers.

CHAPTER 21

Assessment as part of teaching

Here we discuss assessment which is an integral part of teaching, taking place within regular activities. The assumption is that if these activities provide opportunities for children to develop skills, attitudes and ideas which are the aims of their education at a certain stage, then the activities also provide opportunities for these things to be assessed. It is true that assessment has always been part of teaching, for effective interventions and interactions between children and teacher depend on teachers knowing where children are in the development of their ideas and skills. However, it is not always planned so to take advantage of the best opportunities for assessing different aspects of children's development. Thus planning has to be the first consideration, after which we shall look at the nature of the information as it relates to the important matter of using it.

Planning assessment within activities

Collecting information to build up a picture of *all* children in a class over *all* skills, concepts and attitudes, is a formidable task which requires a thought out strategy and careful planning if the result is not to be more information about some children than about others and about only the readily assessed parts of their achievement. The planning with which we are concerned here is not about methods to be used in assessing (which are the subject of Chapters 22 and 23), but the ways of ensuring this full coverage of children and aspects of learning.

Teachers plan specific activities to fit their general plan (which in turn fits into an overall school plan for the curriculum area) and this plan will include the grouping of children, starting points and developments, materials to be used, etc. The consideration of assessment should also be part of this plan, so that the teacher considers the action which (s)he needs to take to collect information which will contribute to building up a picture of each child's progress.

Three points need to be considered in planning for assessment:

(1) which skills, ideas, attitudes will be assessed out of those which could be assessed
(2) which children will be assessed
(3) what part will children play, though self-assessment.

(1) *When to assess what*

Different aspects of achievement offer different assessment opportunities. For instance, many of the skills and attitudes relating to investigation can be assessed in any investigation and therefore the opportunity for their assessment will occur as frequently as children undertake investigations. These can be called 'frequently occurring' aspects. They do not need to be assessed for all children every time they occur, which is fortunate because this would be impossible. By their nature, assessing these skills involves, particularly for young children, careful, on-the-spot observation of how they carry out the activities.

Opportunities for assessing other aspects, however, do not occur so often. Ideas relating to specific subject matter, such as magnetism or seed germination can only be assessed when activities relate to these things. These are 'infrequently occurring' opportunities for assessment. Information needs to be gathered about all the children working on a particular content whilst the opportunity exists. Fortunately these are the aspects of achievement which can be assessed through children's drawing and writing, things which can be studied after the event rather than assessed on the spot, so it is possible to collect information about several groups of children or even a whole class during the time the relevant activities are in progress (See Chapter 22).

In planning what to assess, therefore, it is best to consider

— first, the *infrequently occurring* aspects and make sure that information will be obtained about these for all the children concerned,
— then, the *frequently occurring* aspects, which will be assessed for some of the children.

(2) *Selecting the children*

The greatest benefit of planning assessment is perhaps that it ensures that information is gathered equitably about all the children, not just the ones who need most help or claim most attention. It depends upon keeping records and carrying out the assessment systematically. For frequently occurring aspects, the teacher might plan to observe and make notes about one group particularly during an investigation, which could spread over several sessions. The teacher would not be standing and watching this group for long periods; indeed the special focus in his/her mind should not be apparent to the children. The difference should be in the identification of particular kinds of information which the teacher is gathering during interaction with the group and in the notes (mental and perhaps written) made at the time about each child in the group. This is not in practice as demanding as it may seem in abstract, as an example may show.

This example of work with infants is based on work quoted in Harlen et al (1990). The teacher began the lesson by reading to the whole class a story about feet, in which a boy puts his Wellington boots on the wrong feet. After the story, still as a whole class, they talked about the events in the story, about their own feet, about shoes. This was followed by the organisation of group work, with six groups of four children, and the tasks of each group were described before they split up and went to the tables where necessary equipment had already been put out. So as soon as they reached their tables they could start on something, whilst the teacher began to pass round each group, helping, monitoring and observing.

The teacher had decided that her assessment focus would be group one, which had been given a collection of old plimsolls, Wellington boots, other shoes and slippers to investigate. (Other groups were taking up the subject of feet through different activities, including making shadows of their feet using a torch, measuring them, drawing round them and cutting out the shapes, making a graph, etc.) Group one had the task of discussing which shoes or boots they thought would be most waterproof and so this involved feeling and manipulating the materials. When they had made their choice and given their reasons, they were asked to find a way of showing that their idea was correct. So they planned a simple investigation and proceeded to carry it out. There was plenty of opportunity for the teacher, on her visits to this group to collect information about the detail of the children's observations, their identification of simple differences, their interpretation of the results (water drops stayed on the surface of smooth and hard materials) and the record they made of what they did. At the same time she made notes about individual

The Water dripped of

Figure 21.1

children, for example: 'Lee – said water soaked through the plimsoll tongue because it's thinner than the other part. Tried out drips on sole and rubber edging – predicted that the water wouldn't go through'. Fig. 2.1 shows Lee's record.

As well as the information about the way in which the members of group one went about observing and investigating, the teacher discussed their ideas about materials, particularly in relation to properties which varied among the ones they had been given. In this work on feet the other groups were not working with materials at this time, but the teacher ensured that such opportunities were given through other activities and so enabled her to complete her assessment of all the children's ideas on materials.

An aspect of this approach which may at first seem surprising is that accumulating information over time in this way means that children will be assessed in relation to the same skills, etc whilst engaged on different activities. Does it matter that group one were investigating materials whilst group two might be investigating toy cars when their skills are assessed? Or that ideas about materials were assessed for group one using different materials than other children may be handling when their ideas are assessed? These are questions which need to be answered in terms of the purpose and use of the information, which we will take up in the next section.

(3) *Using the children to help*

The use of children's self-assessment in helping their progress is not well developed in primary schools, apart from the area of language where some materials have built-in means of children testing and recording their achievement of certain skills. Involving children in their assessment means that they must know what are the aims of their learning. Communicating these aims for science is not easy but the rewards of successfully attempting it are quite considerable, not only for help in assessment, but the obvious potential for self-direction in learning. Direct communication of complex learning objectives and criteria of achievement is unlikely to be successful. Certainly there is no point in sharing national curriculum statements of attainment with them! Such things can only be understood by children through examples.

Self-assessment skill has to be developed slowly and in an accepting and supportive atmosphere. It takes time to work through several stages before children are able to apply to their achievement anything like the criteria which their teacher would apply.

The process can begin usefully if children from about the age of eight are encouraged to select their 'best' work and to put this in a folder or bag. Part of the time for 'bagging' should be set aside for the teacher to talk to each child about why certain pieces of work were selected. The criteria which the children are using will become clear. These should be accepted and they may have messages for the teacher. For example if work seems to be selected only on the basis of being 'tidy' and not in terms of content, then perhaps this aspect is being over-emphasised.

At first the discussion should only be to clarify the criteria the children use. 'Tell me what you particularly liked about this piece of work.' Gradually it will be possible to suggest criteria without dictating what the children should be selecting. This can be done through comments on the work. 'That was a very good way of showing your results, I could see at a glance which was best' 'I'm glad you think that was your best investigation because although you didn't get the result you expected, you did it very carefully and made sure that the result was fair.'

Through such an approach as this children may begin to share the understanding of the objectives of their work and will be able to comment usefully on what they have achieved. It then becomes easier to be explicit about further targets and for the children to recognise when they have achieved them.

The nature and use of the information

What kind of information results from gathering it as part of regular teaching? The use of the word 'systematic' to describe it may give a false impression of tidiness and completeness. Reality does not allow this. However carefully a teacher plans to observe and discuss, there are bound to be interruptions and events which mean that practice is not quite as planned. And even when things do go to plan it is inevitable that the information is really of the nature of snippets from listening and watching the activities at intervals. Information from any one lesson is incomplete in relation to anything which would enable a definite conclusion to be reached about what children know or can do. Over a period of time, however, successive pieces of information add together to form a more coherent picture.

Even then, however, the picture is not such a clear one. It will contain contradictory information, as when a child appears to be able to do something in one situation but not in another; it will be incomplete and, of course, always changing. The fact is that this is the character of formative, on-going assessment and we should not think of it as falling short of some more complete and certain information. It is like this because children are real, changing and complex beings and we cannot expect anything other than a certain messy uncertainty about where exactly they are in their development at a particular time.

So when we find that:

> Sarah seemed to be able to grasp a simple pattern in her weather records last week but a similar thing involving patterns in symbols gives her unexpected trouble this week.
> Richard seems to be full of ideas for improving their investigation when working with Jane and George but seemed quite devoid of ideas when Kevin joined them —

we should not feel that our assessment is inadequate but that we have more information than we would have had if we had assessed them (or tested them) in only one situation. We have noted (page 163) the context dependency of children's performance and this becomes very clear when information is collected regularly in a range of situations. It brings the recognition that it is difficult ever to be sure that a child can or cannot do something, and the realisation that test results, which seem to give definite information, do so only because they are too crude to reveal the fluctuating reality of achievement.

In assessment for formative purposes this uncertainty does not matter, since the purpose is not to pin down that a child has mastered this skill or that idea, but to provide a basis for helping learning. The fact that a child can do something in one context but apparently not in another is a positive advantage, since it provides clues to the conditions which seem to favour better performance and those which seem to inhibit it. This is valuable information for taking action.

The use of the information for the purpose of helping individual children is part of the answer to the question of whether it matters that the children's skills and ideas are not assessed on the same activities. There is no comparison being made between children and therefore no need for the subject matter to be controlled providing each activity gives opportunity for the skills and/or ideas to be used. The other part of the answer relates to the nature of the skills and ideas being assessed. It is assumed that these are generally applicable skills, such as those we have discussed in Chapters 4 and 11, and that the ideas are generalised concepts which apply to a range of specific content, not specific facts about certain objects or materials. So again it should be the case that one context is as valid as another for formative assessment if the opportunities are there. As we have said, the variation of performance across contexts is a natural part of performance and information about it adds to the usefulness of the assessment rather than detracting from it.

CHAPTER 22

Assessing scientific ideas

We are dealing separately with the assessment of ideas (in this chapter) and of skills (in the next) only because it is usual for these achievements to be recorded and reported separately. But just as skills and ideas are bound together in teaching so they will be assessed together within the same activities. The kinds of information required are different but can be gathered at the same time, using methods appropriate to each.

This applies to formative assessment for helping decisions about the further experiences for children and to summative assessment carried out by checking up where children are in their development. We will consider and illustrate methods for both purposes in these chapters.

Deciding what ideas to assess

It is worth making explicit what was implied in the discussion in Chapter 21 of planning for assessment. When teachers plan activities and the sorts of experiences they will provide for children they have in mind certain ideas and skills which they intend the activities to foster. These are expressed in various ways (in training courses they may be required as formal statements of the objectives of a lesson) and recently the identification of statements of attainment of the national curriculum have provided a useful means of doing this. It is not expected that a single lesson, or activity spreading over several lessons, will result in the achievement of the skills and ideas but that the activity will make a contribution and this is seen as the educational reason for the activity.

Forward planning of this kind is essential for purposeful teaching and it need not be in any way confining. The stated intention may be that children should raise questions or plan their own investigations or exchange ideas about a new phenomenon; these objectives allow all the freedom children need to retain ownership of their learning and provide a clear basis for a teacher to decide how to set up the lesson and how to interact with the children. It is this clarity which is essential for assessment to take place as part of the teaching, for there must be a clear view of the information which is to be gathered. What this means in practice is best conveyed through an example.

Activities with ice balloons (see PSR number 3, Spring 1987, page 5) provide a spectacular way of enabling children to explore water in its solid form. An ice balloon (created by filling a balloon with water, freezing it and then peeling off the rubber when it is solid) is so much more likely to rivet children's attention than the more accessible ice cubes, that it is worth the effort to make.

Working with eight and nine year olds, a teacher prepared ice balloons and gave one to each group of children with a tank of water in which it could be floated. The children were fascinated and made many observations and raised questions, some of which were followed up on later occasions. Indeed the major set of objectives of the activities was to encourage children to explore, to question and then to set about answering the questions through more systematic investigation. We shall return to these objectives in the next chapter. Here we focus on the ideas that the teacher intended to foster through the activities. These included

- the meaning of 'hot' and 'cold'
- the notion of temperature as a measure of hot and cold
- the idea that heat is needed to melt ice and that cooling causes water to solidify
- the similarities and differences between solids and liquids.

There were a few other ideas which children would no doubt be helped to develop in the activities but the teacher decided to focus on these four rather than spread more widely. They include a range of conceptual difficulty; they span levels 2 to 4 in the national curriculum, which the teacher judged would match the levels of development of children in the class with respect to this area of work.

The question then arises as to how information is to be gathered about the children's ideas. What methods can be used during the course of the activities? In answering this question it has to be recalled that opportunities to assess these ideas may be relatively 'infrequently occurring' (see Chapter 21) and so the methods need to be ones which enable information to be gathered from all the children undertaking the activities.

Methods for use as part of teaching

The main methods fall under the four headings of discussion, including questioning and listening, children's drawings, annotated by themselves or in discussion with the teacher, concept maps, which are a special form of drawing and children's writing, which may be structured by the teacher in the form of questions to answer, or may be an account of observations and ideas structured by the child.

Discussion

Just listening in to the children's observations will provide information about the way they are using the words *hot, cold, melt* and perhaps *temperature*. For those children whose grasp of these is not clear some direct questioning will help to probe their ideas:

— How does the water feel compared with the ice ?
— Touch your arm and then the ice - how does the ice feel in comparison?
— (If the word *temperature* has been used) Which is at the highest temperature, your arm, the ice or the water?
— What do you think it means when the temperature of something goes up?

More open questions help to start discussions about why the ice melts, something which some children of this age will assume is 'natural' and does not need explanation:

— What do you think is causing the ice to melt?
— What would you need to do to stop it melting?
— What do you think is happening when you put water in a freezer and goes solid?

If such questions seem too demanding, remember that, first, they are not much different from the questions teachers ask when they join groups during the normal course of teaching, and second, they do not all need to be asked of all children. It will be evident in many cases that either a child is still some way from an idea or that they have already become at ease with it. This information will be used to lead the children to activities which advance their ideas from where they

The sun is hot
and the water is
cold and the water
sticks to the sun
and then it
goes down
goes down

(Age 7 years)

"The sun is hot and the water is cold and the water sticks to the sun and then it goes down"

Figure 22.1 (Russell and Watt, 1990, p. 29)

are, but there is no need to continue probing when this is clear. A third point is that talking in not the only means of communicating ideas that the children can use.

Children's drawings

Asking children to draw what they think is happening gives a permanent record of their ideas which has the advantage of being able to be perused after the event. Examples of children's drawings which reveal their ideas are given in Chapter 3. It is not easy for anyone to draw abstract things such as ideas about melting, and the use of labels and annotation as a commentary on what is happening is necessary, but the drawing is essential for conveying the image that the child has in mind. For example, the above drawing by a seven year old, Figure 22.1, shows very clearly that the child considered the direct action of the sun as important in causing the disappearance (by evaporation) of water from a tank.

178

In the case of the ice balloon, useful suggestions for probing (and advancing) children's ideas might be:

- Draw what the ice balloon looks like now and what you think it will look like after dinner, at the end of the afternoon and tomorrow if we leave it in the water (the same task could be given to predict what it would look like if left out of the water)
- Put labels on your drawings to point out the things which have changed and what changed them.
- Draw a picture with labels to show all the differences you can between ice and water.

It is best, if at all possible, to talk to the children individually while they are doing their drawings and to clarify for your information things which are not easy to interpret.

Concept maps

Concept maps are diagrammatic ways of representing links between concepts. There are certain rules to apply which are very simple and readily grasped by children of five or six. If we take the words Solid and Liquid we can relate them to each other in this way

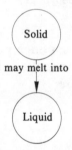

We have a proposition which indicates a relationship, with the arrow signifying the direction. Solids may melt into liquids, but not vice versa. We can add to this by linking other words and so forming a map.

Asking children to draw their ideas about how things are linked up provides insight into the way they envisage how one thing causes another. The starting point is to list words about the topic the children are working on and then ask them to draw arrows and to write 'joining' words on them. Figure 22.2 shows the list and the map which a six year old, Lennie, drew after some activities about heat and its effect on various things. It is possible to spot from this

energy

temperature

degrees Celsius

melt

boil

liquid

solid

friction

evaporate

heat

steam

insulate

food

water

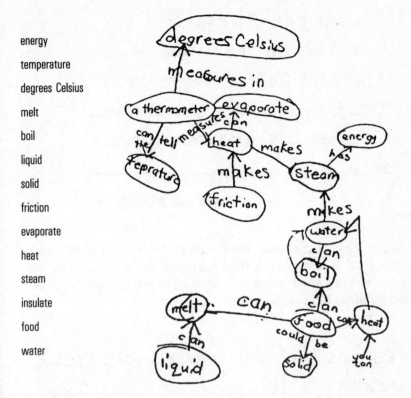

Figure 22.2

that Lennie has not yet distinguished heat from temperature but that he has some useful ideas about what heat can do. As with all diagrams, it is advisable to discuss them with the child to be sure of the meaning intended.

Children's writing

Whilst drawings can usually be made by even the youngest children, writing is most helpful when children become at ease in doing it. However the following example, figure 22.3, was of a six year old who explains why the condensation from her breath on a cold window went away.

I went out side and I
breathed on the windows
and My cold breath comes out
and if you look at it you can see it
go a way it goes when it gets
warm

(Russell and Watt, p. 36)

Figure 22.3

In figure 22.4, a ten year old's answer to how to slow down evaporation of water from a tank indicates the value of not just asking for writing about what has been observed but posing problems where ideas have to be used:

by putting a peice of glass covering it and it will last longer because it can't gas out.

(Russell and Watt, 1990, p. 38)

Figure 22.4

The kind of writing task which may reveal children's ideas in the context of the ice balloon activity would be stimulated by:

• What were the changes which happened to the ice balloon in the water? What would you need to do to stop the ice balloon changing?
• Describe what you think would make the ice melt more quickly and say why it would work.
• Describe to someone who had never seen ice and water what the differences are and how they would know which was which.

Many of the questions suggested earlier for discussion could also be turned into writing tasks for older children.

Using the results

It will probably have been realised that what has been suggested for assessing could equally well have been suggested as teaching points, so close is the relationship between teaching and assessment of the kind being considered. Just to underline further that the purpose of gathering information in this way is to use it in teaching, we bring together here some ideas for acting upon the information about the children's ideas.

(1) If the child is not using a word in a way consistent with a good grasp of the concept, discuss the child's meaning of the word, as suggested on page 80, and try to find out how the misunderstanding has arisen.

(2) The same may apply to the whole task and so it is useful to find out what the child considers (s)he is doing (for example, if melting is understood as 'breaking up' the child's actions could seem quite inappropriate to the teacher but rational to the child.

(3) Use other members of a group to help a child whose ideas seem less well developed. For example the child who didn't see any reason to explain why water freezes in a freezer, was challenged by others to go further: 'But there must be something happening, it doesn't happen by magic'. He then listened to their ideas and began to add some suggestions of his own.

(4) The most effective way of helping children to develop their ideas is to help them to turn them into a form which can be tested. This involves making a prediction. The child who said that it was the water which was making the ice balloon melt revised this after realising that on this basis taking it out of the water should stop it melting, which was soon found not to be true.

Methods for checking up children's ideas

As was discussed in Chapter 20, one of the ways of obtaining information to summarise children's achievement is by checking up, that is, giving some special tasks which are devised specifically to assess the point reached in the development of ideas. There are also times when teachers feel the need to introduce special tasks when

182

it does not seem to have been possible to collect information about certain ideas in any other way. Figures 22.5–22.7 show how this can be done through written questions. They do not need to be given as 'tests' and indeed could seem to the children to be part of their normal activities providing the questions are written to match the subject matter of the activities.

These three examples are from the APU bank of questions and show ways of assessing ideas through asking children to apply them rather that asking them directly for facts which could be memorised. The open response is important so that children can use or explain their ideas and not just respond to alternatives given by others. More information about how children responded to these questions and how they were marked can be found from the references given.

David and John put equal amounts of dry sand, soil, grit and salt in four funnels.
They wanted to find out how much water each one would soak up.   So they poured 100 ml of water into each one.
This worked all right until they came to the salt. When they poured the water in almost all the salt disappeared.

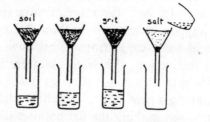

Why do you think the salt disappeared but the other solids did not?
I think this might be because . . . .
. . . .
. . . .
. . . .
. . . .

Figure 22.5

Two blocks of ice the same size as each other were taken out of the freezer at the same time. One was left in a block and the other was crushed up.

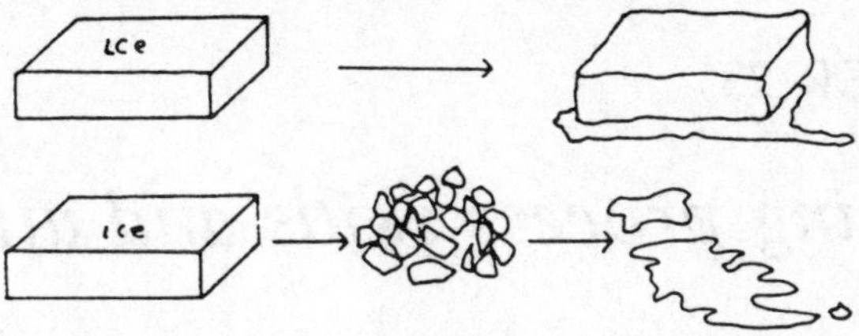

It was noticed that the crushed ice melted more quickly than the block.

Why do you think this was?

I think it was because . . . .

. . . .

. . . .

. . . .

. . . .

Figure 22.6 (DES, 1985, p. 122)

John washed four handkerchiefs and hung them up in different places to dry. He wanted to see if the places made any difference to how quickly they dried.

a) In which of these places do you think the handkerchief would dry quickest? Tick one of these:

☐ In the corridor where it was cool and sheltered

☐ In a warm room by a closed window

☐ In a warm room by an open window

☐ In a cool room by an open window

☐ All the same

b) What is your reason for ticking this one?

. . . .

. . . .

. . . .

. . . .

. . . .

Figure 22.7 (DES, 1984, p. 252)

CHAPTER 23

Assessing process skills and attitudes

In assessing process skills and attitudes we are concerned with aspects of performance which are present to some extent in many science activities, and possibly in any of a practical nature. These are 'frequently occurring' aspects which do not have to be assessed in specific activities and can be spread over a period of time. The general guidance in relation to the national curriculum that the weighting of scientific investigation skills within activities should be around 50% for Key Stages 1 and 2, suggests that opportunities for teaching, and therefore for assessing, these skills should arise as often as science activities occur.

Attitudes, by their nature a part of all behaviour, have to be assessed across a range of activities and so can be treated in the same way as process skills. So there is plenty of opportunity for collecting evidence of skills and attitudes; the difficulty is in gathering the information systematically. The major part of this chapter is concerned with how to do this as part of teaching. There are also examples of using children's written work for checking up on process skills.

What to assess

We need not spend long on this since the range of process skills has been identified in Chapter 4 as

observing, hypothesising, predicting, investigating,
interpreting findings and drawing conclusions, communicating,

and attitudes have been identified in Chapter 5 as

curiosity, respect for evidence, willingness to change ideas,
critical reflection.

Unlike concepts, these stay the same in different science activities, although the emphasis among them varies from subject to subject and with the experience of the children. For, example, the activities of 5 year olds will tend to give lots of opportunities to explore and gain experience; the skills most used will be observing, questioning and communicating. The other skills will be used when the children begin finding ways to answer questions through their activity. Given plenty of previous exploring, six and seven year olds will be covering the whole range of skills in simple activities.

Subject matter, as we have noted in Chapter 21 (p 168) is important in determining the extent to which children can use the skills they have. We don't expect young children to be able to 'process' (that is, use process skills to make sense of) complex phenomena and events, particularly those which can only be understood using mental models, so when their process skills are being assessed they *must* be working on the exploration and investigation of subject matter within their understanding.

Methods for use as part of teaching

The main method for assessing process skills and attitudes as part of teaching (and for helping teaching) is observation during activities. The task seems a formidable one, to make subtle observations about each and every child whilst at the same time managing and fostering their learning! Planning is the key to making this possible, but it also helps to remember that the asessment benefits the teaching – if it doesn't then it something is wrong. There are two aspects to planning in this context: deciding which children to observe and deciding what to observe.

Deciding which children to observe

The observation in any one session or lesson time should be restricted to a small number of children, most conveniently those working in the same group. This will usually be between three and five children. As we have said in Chapter 21, the 'target' groups should not be aware of particular attention being paid to them; *there should be no suggestion of a teacher hovering with a clip board and refusing to interact with the children in a normal way.*

The target group should be selected according to the stage of their activities and the information the teacher wants to collect. The

Children	Observg	Hypothg	Predictg	Investg	Interpg	Commung	Cur	Res	Chg Id	Crit Refl
1										
2	▓	▓	▓		▓	▓	▓	▓		▓
3										
4	▓	▓	▓		▓	▓	▓	▓	▓	▓
5	▓	▓	▓		▓	▓	▓	▓	▓	▓
6										
7										
8	▓	▓	▓	▓	▓					

Figure 23.1

teacher may, for example, be filling in 'holes' in previous observation of these children, holes left when earlier activities did not give the opportunity to observe certain aspects as anticipated. The aim is to have information fitting all the cells in a matrix of skills and attitudes and children such as in figure 23.1. (note that it is *not* a record sheet but a planning device).

In one lesson information within the shaded area may be collected; in another, other children may be observed, or holes filled in, according to the opportunities provided by the activities planned. It may take half a year to complete the information for all the children. Gathering it becomes progressively more easy, however, as the process becomes an established part of planning and interaction with children. Teachers develop a mental framework for assessing children into which information can be fitted straight away without the need to make notes to be interpreted later.

Deciding what to observe

The observation of children is not just a matter of seeing whether or not they are hypothesising, predicting, showing critical reflection, etc. Such broad judgements would be impossible without thinking out what it means to hypothesise, predict and so on. The *indicators* for each process skill and attitude which have been identified in Chapters 4 and 5 are a start. They need to be 'translated' into the subject matter of the particular activities being observed. To provide an example, we return to the work with ice balloons mentioned in Chapter 22 and pick up the process side of these activities.

To give an idea of the experiences of the children, these are some of their observations, recorded by Peter Ovens (PSR, No 3, Spring 1987):

1. There are three types of ice: clear, lined and white, or 'frosted'.
2. The lines are like spiky objects within the ice; they usually radiate out from a point near the centre; they are fairly straight.
3. The ice has air bubbles in it. As the ice melts, they are released and, if the ice is in water, the air bubbles rise and pop at the surface.
4. The ice feels cold, and sometimes dry and sticky, when very cold; the surface is sometimes smooth and slippery, particularly on the clear part, but can feel rough on the frosted part.
5. When the ice is just out of the freezer it feels very cold; a mist can be seen near it, a white frost grows on its surface, and if it is put into water straight away, cracking noises are heard and cracks appear inside it.
6. It feels heavy. It floats in water, but only just.
7. In a tank of water the ice seems to prefer to float the same way up or sometimes in one of two alternative ways.
8. If you leave the ice floating in a tank of water undisturbed, a ridge shape appears at the level of the water; it melts faster below than above the water, and eventually becomes unstable and capsizes. The water level in the tank rises. The water gets colder.
9. The ice is very hard, and difficult to break. When it is broken it makes splinter-shaped pieces.
10. If you put ink on top of the ice, it doesn't soak in, but runs down the side.
11. If you put salt on the top, the ice melts, the salt mixes with the melted ice, and, if it runs down the side of the ice it makes cuts in the surface.
12. There is usually a pale yellow spot near the middle of the ice balloon.
13. Larger pieces of ice slide down a slop more easily than smaller pieces.
14. If you poke the ice with a metal object, a mark is easily made in its surface.

(Primary Science Review, No 3, 1987, p. 5)

During the initial exploration and the follow-up investigations of these kinds of observations the teacher would be able to observe whether or not the individuals within the target group did these things:

Observation

- made detailed observations (of 'lines' or air bubbles in the ice)
- used their senses (to feel the ice, its coldness, its 'stickiness' when just out of the freezer)
- noticed the sequence of events (which parts started melting first)
  - used a simple instrument to aid observation (looked at the ice through a hand lens).

Hypothesising

- suggested why something happens or reasons for observations (why the ice seems to be cracked, why ink runs off and doesn't soak in)
- used previous knowledge (e.g. to explain what happens when salt is put on the ice balloon)
- realised that there were several possible explanations of some things (e.g. several reasons why there were air bubbles in the ice).

Predicting

- using past experience in saying what might happen (e.g. that the ice would melt more quickly in warm water, or in other conditions).

Investigating

- planning to change the independent variable (e.g. placing some ice in air and some in water to test the idea that this makes a difference)
- controlling variables for a fair test (e.g. keeping the initial mass of ice and the temperature the same in the above investigation)
- measuring an appropriate dependent variable (e.g. the change in mass of the pieces of ice over the same time in this investigation).

Interpreting findings and drawing conclusions

- linking pieces of information (eg 'the larger pieces of ice slide down a slope more easily than the smaller ones')
- finding patterns in observations (eg the more pieces an ice cube is broken into the more quickly the whole thing melts)
- showing caution in drawing conclusions (eg 'it isn't everything that melts more quickly in water than in air').

Communicating

- talking and listening to sort out ideas (eg sharing ideas about why a metal object will make a mark in the ice but a wooden one doesn't)
- making notes during an investigation (eg noting the initial mass of the ice cubes put to melt in different conditions; drawing and labelling the places where they are put)
- choosing an appropriate means of communicating to others (eg drawing cubes of diminishing sizes to show how much melted at various times).

Curiosity

- showing interest (by making observations of the kinds reported on page 187)
- asking questions (eg 'why does ice float in water?' 'does it float in every liquid?' 'why does salt make the ice melt?')
- spontaneously using information sources (eg to find out how much of icebergs are below the water).

Respect for evidence

- reporting what actually happens despite contrary preconceptions (eg 'moisture still forms on the outside of the tank when there is a cover on the tank, but I thought it wouldn't')
- querying a conclusion when there is insufficient evidence (e.g. 'we don't *know* the bubbles are air, we just think they are')

Willingness to change ideas

- being prepared to change ideas in the light of evidence ('I thought the moisture on the tank came from the ice, but it can't be that if it happens when the tank is covered').

Critical reflection

- willingness to consider how procedures could be improved (eg 'it would have been better to have had larger pieces of ice to start with, because the difference would have been more obvious')
- considering alternative procedures (eg 'the cube in air ought to have been supported above the table – it was sitting in water half the time'.)

These items are a selection, based on the indicators in Chapters 4 and 5, of those which seem applicable to the ice balloon activities. From these examples it is not too difficult to use other indicators in relation to other activities where they would be appropriate. In the same way the statements of attainment for Scientific Investigation' can be interpreted in the context of particular activities and used in making assessments at the national curriculum levels. Alternatively, observations of the kind above can be mapped onto the national curriculum statements of attainment for the purpose of reporting Teachers' Assessments.

Using children's written work

Not all the evidence of process skills needs to be collected through on the spot observation. Children's written work often gives useful information, particularly if the tasks are set to lead children to describe their observations, predictions, plans and how they carried them out. The examples in Figure 23.2 illustrate the value of the products.

Methods for checking-up process skills

Special practical tasks designed to require all the process skills to be used, to the extent that children are able, have been employed in the APU surveys (DES 1980) and for research purposes (eg Russell and Harlen, 1990). As they require the full attention of an administrator/observer they are not practicable in the classroom. Their value to teachers is in the ideas and hints which they give about the kinds of tasks, ways of presenting them and of questioning children which can be adapted and applied in planning children's practical work.

Also useful for the teacher who wants to check on certain process skills for several children at once are the examples of written questions devised for the APU surveys in Figures 23.3, 23.4, 23.5 and 23.6.

All the APU questions were 'stand alone' items, each being unconnected to any other. The reason for this was the requirement to have a large bank of questions set in different contexts from which to sample for different surveys. Where this is not necessary the questions can be embedded in a theme which is not only more interesting for the children but cuts down on the amount of reading the children have to do to establish a fresh context for each item. An example of

Our prediction is that people will be able to complete the test when they are much closer to the chart and the chart will be not so clear as the first test when they are further away from the chart. We also think that people with glasses will see better than other people because they have more focus in their glass lenses.

If I did this again I would try to think of a way to test the sound and not just guess and try to think of more surfaces and try with different coins at different heights. on the sound I have got two ideas, one, see how far away you can here it drop, and two, get a tape recorder with a sound level indicator.

Figure 23.2

192

When we examined a lychee we found
out that the skin or peel had tiny
hairs on it. When we held it quite
far away the whole fruit looked
like a hard and over grown
rasberry. When we tasted the peel
it was like an advocardo. The peel
was all either red or yellow as I just
said the red tasted like an
advocardo but the yellow was rearly
dicusting this ment that the
fruit is ripe when it is red
or yellow. Then when we took the
peel of tabltaly we found that
there was another skin but this
was transparent. When we took that
skin of we found that the juice
was in some sort of segments like
an orange. Then we tasted the
flesh and it was lovely. After that
we found a stone or seed in
the middle so we cut it open and it
went brown after a few seconds then
we smelt it and it smelt like a
conker (or Horse Chessnut)

(Examples from Vicki Paterson, *PSR*, 1989, p. 17–20)

Figure 23.2 Continued

this approach is given in Schilling et al (1990). Written questions
assessing process skills were devised on the theme of the 'Walled
Garden', which the teacher could introduce as a topic or as a story.
Questions were grouped into seven sections about different things

found in the garden: water, walls, 'minibeasts', leaves, sun-dial, bark and wood. For each section there was a large poster giving additional information and activities and a booklet for children to write their answers. Children worked on the tasks over an extended period, with no time limit; they enjoyed the work which they saw as novel and interesting, in no way feeling that they were being tested. The examples in Figure 23.7 are of the questions on 'minibeasts'. They can be used as guides to setting process-based tasks in other contexts to suit the class activities.

Tom cut an orange into pieces.

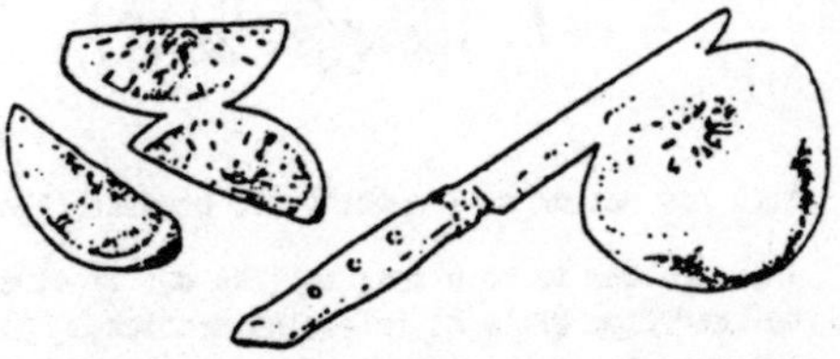

He ate some of the pieces and was going to keep the rest for later.
His mother said: "Cover them with some cling film so they don't dry up".

Tom decided to see if covering them really did make any difference.

He decided to cover some of the pieces of orange and to leave others uncovered. He would see which ones dried up most by weighing them.

To make this a fair test he should make some things the same in case they make a difference to the result.

Write down three things that should be the same.

1. . . . .
   . . . .

2. . . . .
   . . . .

3. . . . .
   . . . .

Figure 23.3

194

Mick and Mary were working on a needlework picture together.

Mick was better at threading the needles than Mary.

They decided to do a test to find out if other boys were quicker than girls at threading needles.
This is what they did:

1. Asked 10 boys and 10 girls to help them
2. Asked each one to thread 5 needles
3. Timed how long each person took to thread the 5 needles.

How would Mick and Mary use these results to find out if the boys were quicker than the girls?

. . . .

. . . .

. . . .

DES, 1985b, p26

Figure 23.4

Some woodlice were put in the middle of a tray containing some wet soil and some dry soil. Half of the tray was then covered with a dark cloth.

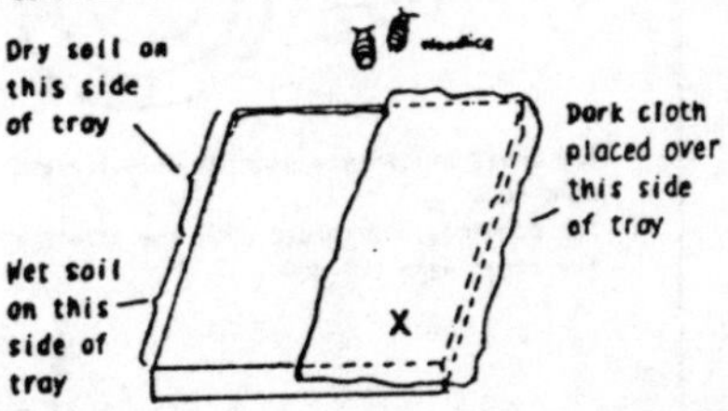

Dry soil on this side of tray

woodlice

Dark cloth placed over this side of tray

Wet soil on this side of tray

X

After 30 minutes all the woodlice were under the dark cloth around the area marked X.

a) Use this information to decide where woodlice are most likely to be.
Tick one of these

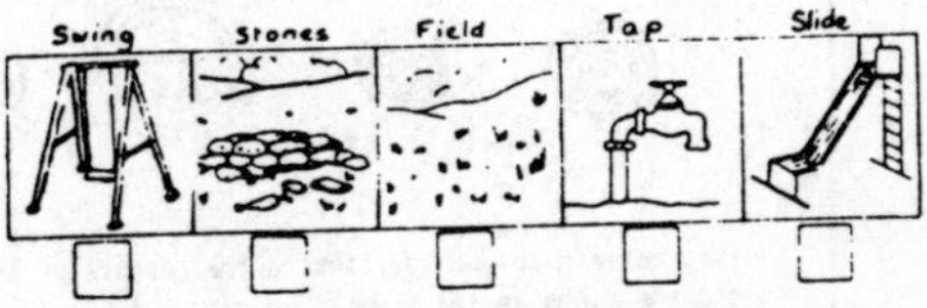

Swing	Stones	Field	Tap	Slide

b) Say why you think this is the most likely place to find woodlice.

Because . . . .
. . . .
. . . .
. . . .
. . . .

(DES 1983, p24)

Figure 23.5

When we cut across the trunk of a tree we see growth rings.

This tree is three years old; it has 3 growth rings

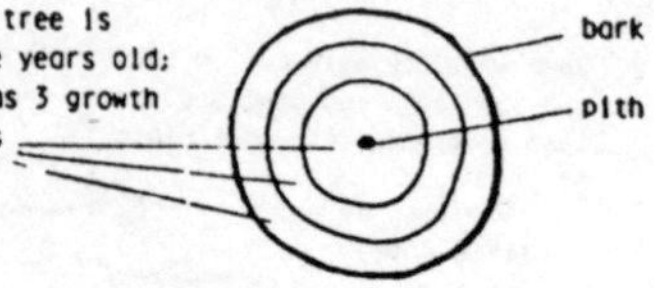

bark

pith

The trees below were planted at different times in the same wood.
The drawings underneath show the growth rings seen when the trees were cut down.

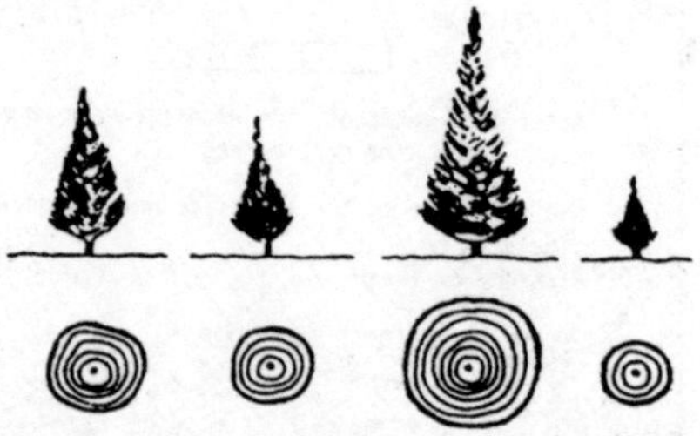

What pattern do you see linking the heights of the trees and the rings in the trunk?

The pattern I see is . . . .
. . . .
. . . .
. . . .
. . . .

(DES, 1983, p25)

Figure 23.6

Minibeasts

Dan and Tammy kept a note of all the "minibeasts" they found in the Walled Garden. They drew the minibeasts as well as they could.

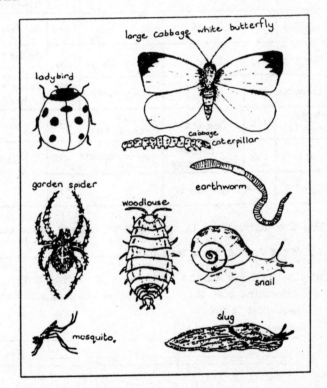

Read about the "minibeasts" in the project folder before you try to answer the questions.

Figure 23.7

Later, back at school, they used some books to get
information about the minibeasts. They made a special
chart, called a table, which showed the information and put
it in the Walled Garden project folder. Here is a copy of it.

Minibeast	legs	where eggs laid	eggs hatch into	sheds skin	adult feeds on
woodlouse	yes	under stones, logs	young woodlice	yes	dead animals and plants
snail	no	soil	young snails	no	dead and living plants
ladybird	yes	plants		yes	live greenfly
slug	no	soil	young slugs	no	dead and living plants
earthworm	no	soil	young worms	no	dead things in the soil
cabbage butterfly	yes	leaves	larva caterpillar	yes	plants
spider	yes	in cocoon on leaves	young spiders	yes	flies
mosquito	yes	on water		yes	

1. Use the information in the table to answer these questions:

   a) What do ladybirds feed on? ....

   b) In the table all the minibeasts with legs have
      something else that is the same about them. Can you
      see what it is?

      ....

      ....

2. When they made the table they could not find all the
   information about the ladybird and the mosquito.

   Please fill in this information for them on their table:

   a) A ladybird's egg hatches into a LARVA.

   b) Adult mosquitos feed on ANIMALS and PLANTS.

Figure 23.7 Continued

3. Dan and Tammy's table shows that snails eat dead and living plants, but it doesn't say whether they like to eat some plants more than others.

Suppose you have these foods that snails will eat:

strawberries   porridge oats   lettuce   carrot

and as many snails as you want. Think about what you would do to find out which of these foods the snails liked best.

a) Say what you would do to start with? (Draw a picture if it will help.)

b) Say how you will make sure that each food has a fair chance of being chosen:

   ....

   ....

   ....

c) What will you look for to decide which food was liked best?

   ....

   ....

   ....

Figure 23.7 Continued

4. What other things could you find out about snails by doing investigations with them?

    Write down as many things as you can think of to investigate.

    ....

    ....

    ....

    ....

5. Dan and Tammy went to visit their Aunt and looked for minibeasts in her garden. They found them all except for snails although they looked carefully for a long time.

    a) Write down any reasons you can think of to explain why there were no snails in their Aunt's garden.

        ....

        ....

        ....

    b) Their Aunt thought it could be because of the kind of soil where she lived; there was no chalk or limestone in it.

        What is the main difference between snails and other minibeasts which Dan and Tammy found?

        ....

        ....

        ....

    c) Why do you think snails only live where there is chalk or limestone in the soil?

        ....

        ....

        ....

Figure 23.7 Continued

CHAPTER 24

Recording Experiences and Progress

There are very many varieties of records which have been proposed for different purposes; teachers in England and Wales can obtain advice and examples from the NCC's Non-Statutory Guidance and from SEAC, from the SCCC and so ED in Scotland, as well as from fellow teachers through the publications of the subject associations (Primary Science Review and the joint publications of the ASE, NATE, ATM and MA) and guidelines from LEAs. This chapter does not attempt to review all of this material, or to compete with it, but to provide a discussion of the kinds of records which may be useful for three purposes:

- recording activities and experiences
- on-going records for formative assessment during each term or year
- summative records which show progress over several years.

All these are records relating to individual children, as distinct from the planning records, mentioned in Chapter 19.

Recording systems are so important to teaching that they have to be a part of it, designed to suit a particular way of organising and teaching the curriculum. Therefore the most useful and least obtrusive ones will be those which teachers devise for themselves having considered the pros and cons of the ones suggested in various publications.

Records of the activities of individual children

Unless all the children in a class are always engaged in the same activities as each other, there is a need for a system which records what individuals have actually been engaged upon. Indeed, even if the activities *were* the same for all it would be no guarantee that their *experiences* would be identical, since children attend selectively to different parts of the work, extend some and give scant attention to others. This is a useful reminder that we can never know exactly what each child has experienced, and in order not to give the false impression that we can it is perhaps better to record activities than to pretend to be more precise.

The purpose of a record of activities is to enable the teacher to ensure that all children, at various times have the learning opportunities planned in each term, in so far as this is possible, and to record gaps in children's experience where this has not been possible. A handleable recording system must be capable of keeping track of children's activities when many different kinds may be on the go at once. For example, if we recall the infant activities about feet described in Chapter 20 (p. 169), the six groups were working on activities which, although having the common theme of 'feet', had quite different learning outcomes (some relating to materials, some to shadows, some to measurement, etc). In this case the teacher decided not to rotate the groups of children round all the activities, so at the end of the topic some children would have had more experience of materials and some more experience of investigating shadows, than others. The teacher would need to ensure that future activities would equalise children's learning opportunities. There are very many different ways of giving the same learning opportunities and these will usually be part of teachers' planning.

It is helpful to distinguish between topics (such as 'feet' or 'movement'), which are likely to encompass activities in various areas of the curriculum, and activities within them. Activities can be identified in subject based terms without disturbing cross curricular work, since they can be regarded as constructs for the teacher to use in planning and analysing the children's work. By listing both topics and activities, a record such as figure 24.1 indicates the context in which the activities were carried out. Such a record could include all activities, but it would be less cumbersome if different ones were kept for the science, language, mathematics, etc., activities within the topics.

The record would be completed by ticking the activities undertaken by each child, not an arduous matter. In many cases groups of children will have undertaken the same activities but the individual record means that absences and changes of group composition are taken into account. Certain activities will probably be regarded as equivalent to each other, whilst in other cases it may be that the context is so different that repetition is desirable. Taking these things

Record of science activies.   Term: ....   Class: ....

Topics Dates	1				2				3					
Activities brief description														
Children														
Belle Adams	✔	✔			✔					✔				
John Allan		✔	✔			✔		✔						

Figure 24.1

into account, the teacher will use the record to keep an eye on the gaps in the activities of individual children and act on this, either in planning the next term's work or having one or two sessions in which children are directed to activities which they have missed.

It would be possible to add a row to this record which indicates the statements of attainment and perhaps the strands to which each activity relates. This would have value in showing when children had been involved in activities which could lead to achievement of the statements of attainment, but it would not, and should not, be assumed to indicate anything about achievement. It could be argued, indeed, that it would be better to avoid any possibility of this assumption by not including statements of attainment, assuming that the planning of the topics and activities had taken into account the programme of study and the statements of attainment appropriate to the overall school programme and the children's previous achievement.

On-going records of children's achievement

These are the most detailed of the records which have to be kept and it is inevitable that a fairly lengthy recording sheet will have to be used for each pupil. A compromise has to be struck, however, between a scheme which is so burdensome that it will not be maintained and is too detailed to be easily used and one which

Record of achievement in science for: Belle Adams:

Attainment targets	level 1	level 2	level 3	level 4
AT 1	a	a b c	a b c	a b
AT 2	a b c	a b c d	a b c d	a b c d
AT 3 etc.	a b	a b c d	a b c	a b c d

Figure 24.2

consists of ticks or other marks whose meaning is rather broad. In deciding which kind of record is most useful it should be noted that:

(1) The over-riding purpose is to help the teacher remember where each child has reached in development so that suitable activities and encouragement can be given.
(2) These records are for the teacher's own use and so the level of detail can be adjusted to suit individual's ways of working.
(3) They will be summarised for other purposes, for school records passed from class to class and for reporting to parents.

Teachers vary as to how much information they can carry in their head and how much they like to write down and this may be one of the factors which leads to a preference for the less detailed form of Figure 24.2 or for Figure 24.3 which gives more opportunity for comments, caveats and explanations.

In figure 24.2 the record is made by circling (or highlighting) the letter corresponding to the statement of attainment which has been achieved. Use of different coloured pens on different dates would enable a time dimension to be built in.

Both examples in Figure 24.2 and Figure 24.3 can be adapted to cover different ranges of levels, according to the age and achievement of the children. Both could also be part of a whole curriculum record through the addition of statements of attainment for other subjects.

Record of Achievement in Science for: Belle Adams

Statement of attainment	Date	Comment	Decision
AT 1			
1a			
2a			
2b			
2c			
3a			
3b			
3c			
AT 2			
1a			
1b			
1c			
2a			
2b			
3a			
3b			
3c			
3d			
etc.			

Figure 24.3

The apparent burden of keeping these records can be kept in perspective by recalling that:

(1) The levels of achievement are coarse grained; on average children will progress one level every two years and so they will not be hopping from one to another without time for the teacher to check on the assessment and record it.
(2) A limited number of attainment targets will cover the work during any particular recording period.

Many teachers will feel that the problem of these records is not that they contain too much information but that they contain too little. The richness and complexity of children's performance can rarely be captured in a brief note, far less by a tick. In many systems, therefore, these records are only a part of the material which is available to a teacher about individual children. Samples of work, chosen by the child and/or the teacher, and more extensive notes may be kept in a file for each child.

The letters a, b, c and so on, represent the statements of attainment for a particular level and attainment target. They have to be 'translated' into the criteria in the curriculum. In some records which have been suggested (e.g. Emery et al, 1990) these are spelled out so that the record is more self-contained but this of course makes for a much longer document. Personal preference is the deciding factor.

Summative records

Here we are considering summative records which are created by reviewing and summing up on-going records rather than the results of checking-up tests. However, if test results exist for certain aspects of performance, then they can be added to the record in a way which clearly indicates their origin.

Two purposes of summative records can be distinguished: those for keeping a record within the school of the progress of individual pupils (cumulative records) and those which are intended for reporting at a particular time to well defined audiences, mainly parents and teachers who will be receiving the children for the next school year.

For cumulative records the same structure can be used as in the on-going records from which it is derived, that is, in terms of the statements of attainment which have been achieved. Figure 24.4 is one example based on this approach. It is essentially a list of statements of attainment, photo-reduced, which are highlighted, using a different colour at the end of each year.

However, Figure 24.4 may be thought to be too detailed to be efficient for the purpose of giving an overview of progress; records at the level of attainment targets only may meet this need better. Too much detail is as inefficient as too little if it obscures the message that the information is intended to convey. Figure 24.5 shows an alternative which can be adapted to termly or annual entries.

These could be science-only records or part of a full profile across the curriculum.

For records which aim to give a picture of where a child is at a certain point, as opposed to a history of progress up to that point, a mixture of achieved levels and open comment may well be required. For different reasons the profile component level (or strand) may be the best for reporting to parents and to receiving teachers.

Parents may find the detail of attainment targets too great to answer their main need to know 'how is (s)he doing in science?' Remembering that parents will have to be taking in information about nine areas of the curriculum, there is clearly a limit to how useful it is to subdivide each of these to any extent. The open

Attainment Target 4: Materials and their behaviour
Pupils should:

1. a) be able to identify familiar and unfamiliar objects in terms of simple properties.

2. a) be able to group materials according to their physical characteristics.
   b) know that heating and cooling everyday materials can cause them to melt or solidify or change permanently.

3. a) know that some materials occur naturally while many are made from raw materials.

4. a) be able to compare and classify materials as solids, liquids and gases on the basis of simple properties which relate to their everyday uses.
   b) know that materials from a variety of sources can be converted into new and useful products by chemical reactions.

Attainment Target 5: Energy and its effects
Pupils should:

1. a) understand that things can be moved by pushing or pulling them
   b) be able to describe how a toy with a simple mechanism, which moves and stores energy, works
   c) understand that magnets can produce pushes and pulls
   d) know about the simple properties of sound and light, including loud/soft, bright/dark, high/low notes and colours.

2. a) understand that pushes, pulls and squeezes can make things start moving, speed up, swerve, stop or change shape
   b) know that some materials conduct electricity well while others do not
   c) know that light passes through some materials and not others and that when it does not shadows are formed

3. a) know about the factors which cause objects to float or sink in water
   b) understand how energy is used and transferred in models and toys
   c) know that a complete circuit is needed for electrical devices to work
   d) know that light, or sound, can be reflected.

4. a) understand that the changes in movement of an object depend on the size and direction of the forces acting on it
   b) be able to construct simple electrical circuits from diagrams
   c) know that sound travels at a different speed from light
   d) know that light travels in straight lines.

Attainment Target 4: Materials and their behaviour
Pupils should:

Attainment Target 5: Energy and its effects
Pupils should:

Figure 24.4

Science achievement record   Name: .... Date of entry to school: ....

Term	Year 1			Year 2			Year 3			Year 4			Year 5			Year 6		
	1	2	3	1	2	3	1	2	3	1	2	3	1	2	3	1	2	3
AT 1 level 1																		
2																		
3																		
4																		
5																		
AT 2 level 1																		
3																		
4																		
5																		
etc																		

Figure 24.5

Scientific investigation

   This includes the skills of carrying out science investigations. Some are practical skills but most are concerned with planning investigations, with careful observation and measurement and other ways of collecting evidence, interpreting findings and drawing conclusions. Also included are communicating results in a suitable way and discussing what has been done in order to improve later investigations.

Overall level of achievement: Level ....

Comment:

....

Knowledge and understanding of life and living processes

This section concerns the growing scientific knowledge of the variety of living things and the processes which characterise living things. It includes knowledge of how living things, including humans, interact with and affect their environment. In the primary school this knowledge comes largely from the children's own exploration and investigation of things around them.

Overal level of achievement: Level ....

Comments: ....

Figure 24.6

comment in addition is important to help summarise and interpret the meaning of the 'levels' achieved and to add other information about attitudes and effort and extra-curricular activities.

Receiving teachers may also find it difficult to use detailed information, not because the meaning is not understood, but because they have the records for a whole class to take on board. So again, profile component levels may be most useful, simply because they are used. Additional information is important, also, but will be of a different kind from that given to parents, pointing out the particular help that a child may need or the kinds of encouragement to which (s)he responds. More detailed information, for reference if required could be provided by supplying, in addition, a copy of the cumulative attainment target record as in Figure 24.5

Figure 24.6 indicates the way in which information may be provided for parents, and with some modification, for receiving teachers. It will be most useful if it is one page of a report booklet, which can include a return slip for parents to make a response to the report.

CHAPTER 25

Indicators for Evaluating School Science Provision

The requirement for schools to report on their own performance to their school governors or board is one reason for being concerned with self-evaluation, but it is not the only reason. More important is the need for heads and teachers to keep track of what is happening in a school so as to identify areas which need attention and plan for development in an ordered way. In order to do this it is necessary to think about what aspects of performance are to be monitored and what criteria used in evaluating them. This is where performance indicators come in.

What are performance indicators?

Suppose you are asked to describe the secondary school you attended to someone who wants a general idea of what the school was like. Your answer might include comments such as:

- good examination results
- old fashioned discipline
- mixture of different types of buildings, well spaced out
- kept parents at a distance
- traditional organisation and teaching.

These are *evaluative* remarks. Some *judgements* have been made about certain features and it is not difficult to separate the kinds of information used from the judgements made on it. Remarks are made about:

- examination performance
- pupils' behaviour and discipline
- age and type of buildings
- relationship with parents
- organisation and teaching style.

Had the school been different then different judgements could have been made about each of these things. The same criteria would have been used to make the judgments. These criteria are examples of performance indicators. Arriving at an evaluative statement, or judgement, involves both evidence and criteria –

Judgement   = evidence + criteria
(evaluation)   (performance indicators)

(eg 'old fashioned   =   ?   + (eg 'pupil behaviour
discipline')   and discipline')

We shall mention methods of collecting evidence later in this chapter.

It is clear in this example that the choice of performance indicators was fairly arbitrary, since many other things could have been included (expertise of teachers, friendliness of the atmosphere). This illustrates the first of several features of performance indicators which should be recognised:

There is no agreed set of performance indicators for a particular purpose. Educators argue fiercely, for example, over the extent to which pupil achievement in examinations or tests should be used as an indicator of a school's effectiveness.

A second point is that performance indicators are defined in broad or detailed terms according to the purpose for which they are being used. If the description of the secondary school had been requested by someone who was thinking of applying for a post at the school then the description above would have been very inadequate. They would have wanted detailed accounts of the school organisation and curriculum, management style, staffing levels, etc. Such detail would, however, have been inappropriate for the general enquirer, who would have become lost in the mass of facts and unable to grasp the overall picture which was wanted. Thus: *Performance indicators should be selected according to the purpose.*

Statements at a general level are sometime described as 'first line' performance indicators. They are often sufficient for giving an account to people who wish to know about how a school is performing but are not concerned with taking any action on the basis of the information. They need to be broken into more detailed aspects to be useful for the latter case. Although performance indicators vary with the purpose and can be chosen by anyone, relevance to all schools is an obvious advantage. This is particularly so if they are used to compare one institution with another. So, *Performance indicators should be generally applicable across institutions of the same kind.*

It should, of course, go without saying that they should be useful. For purposes where action is to be taken this means that they should relate to features of current concern, and in a timescale where decisions could be made on the basis of the information. Usefulness also means providing feedback to those wishing to take action particularly about the possible effects of certain kinds of action. In essence: *Performance indicators should provide timely and predictive feedback which can be used by those concerned.*

In summary, then, performance indicators describe the kinds of information which can be collected to monitor the performance of an institution,or, more generally, of any system. There are used by industry and organisations which provide services. The waiting list of patients is, for example, an indicator used in the health service to evaluate performance of hospitals. The industrial link often creates concern that simplistic indicators will be applied in a complex system such as education. Schools cannot adequately be evaluated in terms solely of some kind of 'output'. However, if the desirable features just listed are adhered to then it is more likely that appropriate indicators will be used.

Using performance indicators for science at the school level

A report on the state of science teaching and learning in a primary school would refer to information about:

> The time spent on science in each class
> The existence and content of a school policy (relationship to national curriculum guidelines)
> The relationship of class programmes to an overall school plan
> The amount and type of equipment and other resources
> The support and advice to staff
> The performance of pupils

At the class level more detailed performance indicators, relating to the opportunities for developing skills, attitudes and concepts, teaching methods and activities would be appropriate. These will be discussed in the next chapter. At each level the information gathered should be that which can be used at that level. Thus it is relevant to make judgements at the school level about matters of policy but not about the particular activities of pupils which are decided by individual teachers.

Using performance indicators means gathering information relating to them. This information will, in general, be a mixture of *quantitative* and *qualitative* data. Quantitative information, answering the questions:

How many? How often? How much time?

is often described as 'hard' information as it seems to be more objective and reliable than qualitative information which has to be concerned with opinion about

How well? How useful? How appropriate?

However, quantitative information is only useful when it is compared with an amount which is acceptable. So, finding that classes spend 5% of time on science is hard information but has to be related to what is thought to be the optimum time. If this is 10% then the action it leads to is different than if the expectation is actually 5%. So, even though the quantity is known, there still has to be some judgement applied to decide whether or not it is adequate.

The performance indicators listed above for science at the school level can be divided into quantitative and qualitative criteria as follows:

Quantitative:
Time spent on science in each class
Quantity of different kinds of equipment/resources
Time spent on staff support
Numbers of pupils at different levels of achievement
Numbers of pupils enjoying science

Qualitative
Existence of a school policy
Extent to which content reflects national guidelines
Relationship of class programmes to school plan
Suitability of equipment and resources
Adequacy of support/advice available to staff.

Note that there is some refinement in the statements made necessary by thinking about what kind of information should be gathered. Indicators which are a combination of two kinds of information would be difficult to use in practice and so this step of further refining makes them more useful.

Now we shall consider two uses of performance indicators: in on-going review of the school and in monitoring changes which are planned and introduced.

School audit

One use of performance indicators implied in the discussion so far is for on-going self-evaluation of the school. The word 'audit' has come into use for this process. School audits are reviews of strengths and weaknesses which cover all aspects, from buildings and resources to curriculum provision and relationship with parents. They can be carried out internally, by the staff or externally, by local or national inspectors. There is usually some auditing every year (curriculum provision, staff development, resources, pupils' achievements, for instance); but not all aspects will be included every year.

On the basis of the review (audit) and in the light of other constraints (such as the phasing in of new curriculum guidelines for a subject area), the school will decide its priorities for action. For example, it may not be difficult to decide that science needs to be considered a priority if the audit shows that children spend too little time on science, or that the school policy has collected dust or that staff feel under-resourced and lacking in confidence for teaching science. The audit will show just where the weaknesses are and therefore the focus of the action plan.

Action plans

Once an area, such as the development of a policy for science, has been identified, the general objectives are identified. It may be that it is only part of the policy document which needs attention, or that the policy document is sound but is not being implemented. The particular emphasis will be reflected in the overall objectives of the development. Then it is time to get down to thinking out what should be done in relation to these objectives. This is where the school staff have to work out what to do, who will help to do it and within what timescale.

As an example, take a school which decides that the whole area of science work needs to be replanned because evidence in relation to the indicators above gave a generally poor picture. They considered that too little time was spent on science, planning was uncoordinated, staff avoided science activities and pupils were not having the opportunity to achieve the skills and knowledge which were expected of them. The staff might benefit from the help of an adviser during a brainstorming session (or two) as a result of which they might formulate their policy in terms of the following targets:

- To ensure that children spend no less than 10% of their time on science as part of other activities or separately.
- To agree annually on a coherent school plan of science-related topics for each class.
- To set up a central store of equipment and teaching materials.
- To provide in-service in science for staff who need it.
- To ensure that all pupils have opportunities to develop science skills, attitudes and ideas as indicated in national guidelines.
- To improve children's performance in science.

These targets would not necessarily be supported by other schools. There are differences of opinion, for example, about the centralisation of resources. However, the point here is that they are practical, and specific enough to provide a basis for action. The planning needs to be continued so that responsibilities for implementation and deadlines are agreed.

An important part of the plan is the inclusion of the basis on which success is to be judged. This is where the performance indicators come in. In this case they are likely to be very similar to the quantitative and qualitative statements identified above, although performance indicators often have to be redefined as part of the action planning to suit the targets identified.

Methods of collecting information

It is important that methods used to collect evidence for the evaluation by staff in schools should be easy to use, not too time-consuming and should fit in with the school's practices. Methods which require training, intricate procedures, are intrusive or take up a great deal of time cannot be considered however reliable they may be. In the balance between precision and convenience, the latter must win. Thus the methods outlined here are simple and straightforward.

As they are mostly self-evident, they do not require much explanation.

Collection of facts and figures

This means keeping records and bringing together the findings in relation to the quantitative performance indicators. For example:

- asking teachers to log the time spent on science and then collecting the information together
- cataloguing resources and equipment and summarising
- keeping a record of in-service sessions and other staff development activities
- collating teachers' records of pupil performance
- devising or obtaining some means of measuring children's liking for science work.

Documentary analysis

This means carefully reading written statements and noting the attention they give to certain aspects of their subject. It is useful, for example, in evaluating the extent to which a school programme for science is consistent with national guidelines. Some scheme for counting certain kinds of statements or references to certain skills, topics, etc can easily be devised and then both documents analysed using it. The relationship of the attention given in one document to that given in the other is a guide to the extent of consistency between them.

Opinion seeking

People can be asked for their opinions or judgements on certain matters. Those asked must be knowledgeable about the matter in hand and their views should bear weight with the school staff. Often it is the teachers themselves whose opinions are sought. In other cases it might the views of an adviser, or in some matters, of parents or pupils. Opinions can be sought by interview or by questionnaire. The choice will obviously depend on the number of people to be consulted and their accessibility. For example, an adviser's view might be sought on the acceptability of the extent to which the schools' programme reflects the national guidelines (the documentary analysis will show the extent of the agreement but not whether this is within an acceptable margin). In a large school a questionnaire might be used to elicit teachers' views on the suitability of the equipment and teaching materials held by the school. Staff involved in in-service might be interviewed to find how satisfied they are with the help they have been given.

Observing practice

In the case of the science plan this is likely to mean classroom observation although in other cases it could mean observing staff meetings, parents' evenings or pupils in their free time. Classroom observation is extremely useful for a variety of purposes but is, of course, time consuming. If two teachers are able to be in a class together, one can observe pupils or the teacher at work, using either a structured or unstructured method as agreed. These observations might be directed at providing the information about the opportunities children have for using and developing scientific skills and attitudes, for example. (For more information about classroom observation methods see Cavendish et al, 1990, Hopkins, 1985).

CHAPTER 26

Indicators for Teachers' Self-Evaluation

In this chapter the theme of self-evaluation is taken into the classroom, the purpose being, as before, not to make judgements about degrees of success but to use judgements to improve the quality of science education provided for the children. The whole of this book has been about learning experiences for children which might seem to be an unrealistic 'ideal', not readily attainable – not all at once, anyway. It is not expected that it is, but making changes towards what we wish to achieve means being aware of where we are now. Development in teaching is cyclic:

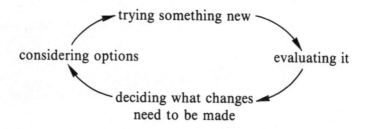

'Evaluating it' is a pivotal step in this cycle. There are several meanings for 'it', including the materials and activities for which evaluation criteria were discussed in Chapter 8. We concentrate here on the two central foci for evaluation in the classroom; the children and the teacher. In each case performance indicators and ways of collecting and using information are discussed.

Focus on the children

Performance indicators

The following aspects of children's activity are suggested as a basis for judging its value for learning. The extent to which children are:

- spending a high proportion of their time 'on task', talking to each other about their work, being busy with it;
- absorbed in their work, finding it important to them;
- understanding what they are doing, not just following others;
- working at an appropriate level so that their ideas are being used and developed;
- handling or investigating materials to answer their questions;
- using thinking and manipulative skills effectively in advancing their ideas.

Gathering and using information

These performance indicators are deliberately small in number and large in significance. They are chosen not to cover every aspect of children's actions and responses but to identify some of the most important dimensions. Information of several different kinds needs to be collected and brought together to make a judgement about each one. For example, the extent to which a child understands what (s)he is doing will be judged on the basis of the way the child talks about the work to the teacher and to other children, what the child writes or draws about it and how the child refers back to it subsequently.

The teacher will need to draw upon all of this information; it is not a matter of making one set of observations and putting a tick on a list, but of reflecting on all that is known. Doing this for all the children and deciding the proportion of the class for which each indicators can be said to be met satisfactorily clearly takes time. This helps to explain the preference for a small number of indicators. But although it does seem that a great deal of information has to be brought together, the saving grace is that the same sources provide information for several indicators at the same time. The technique is to have the performance indicators in mind and to 'comb through' the information from observing children, talking with them, reading their work which is collected as during the course of teaching.

Working with another teacher on this has great benefits. Two teachers wrote about this in the context of ways of observing their children:

We have found that by far the best system is to have an observer in the room who is free to observe while the teacher is able to teach and administer. After the science sessions we have 'talk-back' sessions which are mutually profitable. We are always amazed to find out what we have missed [as the teacher] during the session. Sometimes certain individuals appear with insight we would never have thought them capable of. It really does let us get to know our children even better.

(Jameson and Adams, *PSR*, Summer 1989, p. 9)

Using the information means deciding what changes to make. Some of the possible options for this are described in the earlier chapters to which reference is made in the following.

Children spend a high proportion of their time on task and show absorption in their work when they feel they have ownership of what they are doing and find it interesting. If many children are not showing these signs of interest it may be that not enough time has been spent 'setting the scene' and giving opportunity for free exploration before focussing on specific tasks. It may also be that children could be more involved in defining their tasks and planning how to set about them (see Chapter 10).

If children are not understanding what they are doing this could be symptomatic of some problem of 'matching'. A further look at Chapter 9 may help. Mismatching can be through activities making too low as well as too high a level of demand. To avoid this, more information may be needed about children's initial ideas (Chapters 3 and 10).

For children who seem reluctant to interact with materials and engage in investigation the first thing to check is that sufficient materials are available and accessible to every child and that there is obvious encouragement for children to handle what is there (Chapters 5 and 12). If opportunities exist, the ideas for developing children's process skills in Chapter 4 and 11 may help.

Focus on the teacher

Evaluation of the children's actions and responses must, of course, have implications for the teacher. It may be that changes are needed in the nature of activities provided, the range and amount of equipment and other resources made available or in the kind of help that the teacher gives. But there is also room for self-evaluation in relation to other aspects of the teacher's role suggested by performance indicators such as the following.

Performance indicators for self-evaluation

Teachers' perception of the extent of their:

- understanding of the scientific ideas which the children's activities involve;
- ability to respond to the children's questions;
- provision of necessary and suitable materials and equipment at the time they are required;
- awareness of the children's ideas about the topic being studied;
- knowledge of the development of the children's process skills and attitudes and of procedures for developing these;
- interest in the topic and in the children's response to it.

Gathering and using information

All of the information needed for making judgements in relation to these performance indicators is available through reflection on the experience of planning and conducting science work with the children. What is required in addition is the motivation to put questions to oneself honestly and to face the answers. The process is unhelpful only if nothing is done about what is found. Help with the matter of children's questions can be found in chapter 16 and Chapter 22 suggests ways of finding information about children's ideas. Chapters 11, 12 and 23 assist in getting information about, and helping children to develop skills and attitudes.

The outstanding matter concerns the teachers' own understanding of and interest in the scientific ideas involved in the children's activities.

Teachers' scientific knowledge

There are several relevant and apparently contradictory comments to make about this subject. The facts, according to HMI, are that 'the most severe obstacle to the improvement of science in the primary school is that many existing teachers lack a working knowledge of elementary science appropriate to children of this age' (DES, 1978). It is doubtful whether the situation would be judged to be better now. Indeed recent evidence of teachers' scientific knowledge gathered by a research project has shown several areas where this is judged to be inadequate (e.g. Kruger and Summers, 1989).

Despite these facts there is a strong opinion that many teachers actually know more of what is really relevant to teaching primary science than they think they know. The point here is that a great deal of worry about 'not knowing enough' results from misunderstanding of what teaching primary science involves. If it is seen as the transmission of information (as experienced in their own schooling) then it is understandable that teachers feel very unprepared if they do not have the information to transmit. But the main message of our earlier chapters has been that teaching science to children means enabling them to engage in scientific exploration and through this to develop their understanding. There is no short cut to this understanding through a quick fix of facts.

At the same time this is not an argument for saying that teachers do not need understanding themselves; in fact quite the opposite, for without this they are not in a good position to guide children to materials and activities which develop their understanding. But the emphasis is not on facts but on the broad principles which, as adults with much existing relevant experience to bring together, teachers very quickly grasp, and, most importantly, on the understanding of what it is to be scientific.

This brings us back to the point at which this book began, the nature of scientific activity. It is, therefore, a suitable point to end, but not without giving the last word to someone whose view is surely to be respected, and a message to be cherished:

Science is nothing more than a refinement of everyday thinking.

Albert Einstein

References

Association for Science Education (ASE) (1990) *Be Safe*. Hatfield: ASE.

ASE/ATM/MA/NATE, (1990) *The National Curriculum: Making it Work for the Primary School*. Hatfield: ASE.

Barnes, D. (1976) *From Communication to Curriculum*. Harmondsworth: Penguin.

Bell, B. and Brook, A. (1984) *Aspects of Secondary Students' Understanding of the Particulate Nature of Matter*. Leeds University.

Brook, A. Briggs, H. and Driver, R. (1884) *Aspects of Secondary Students' Understanding of Plant Nutrition*. Leeds University.

Catherall, E.A, and Holt, P.N. (1963) *Working with Light*. London: Bailey Bros and Swinfen Ltd.

Cavendish, S. Galton, M., Hargreaves, L., Harlen, W. (1990) *Assessing Science in the Primary Classroom: Observing Activities*. London: Paul Chapman Publishing.

Central Advisory Committee on Education (CACE) (1967) *Children and Their Primary Schools*. London: HMSO

Children's Learning in Science Project (1987) Information Leaflet. Centre for Studies in Science and Mathematics Education, University of Leeds.

Clough, D. (1987) 'Word processing in the classroom and science education', *Primary Science Review*, No. 5, p. 5.

Conner, C. et al (1991) *Assessment and Testing in the Primary Science Review*, No. 5, p. 5.

Davis, B. and Robards, J.A. (1989) 'Progression in primary science: the assessment of topic work', *Primary Science Review*, No. 9, pp. 7–8.

Davis, S., (1989) 'Implementing the national curriculum', *Primary Science Review*, No. 11, pp. 19–20.

DES (1978) *Primary Education in England*. London: HMSO.

DES (1981) *Science in Schools: Age 11 Report No 1.* London: HMSO.

Department of Education and Science (1983) *Science at Age 11. APU Science Report for Teachers No 1.* London: DES.

DES (1984) *Science at Age 11. APU Report No 3.* London: DES.

DES (1985a) *Science in Schools: Age 11 Report No 4.* London: DES.

DES (1985b) *Sample Questions in Science at Age 11.* London: DES.

DES (1988a) *Education Reform Act.* London: HMSO.

DES (1988b) *National Curriculum: Science for Ages 5 to 16.* London: DES and Welsh Office.

DES (1988c) *A Report. National Curriculum Task Group on Assessment and Testing.* London: DES and Welsh Office.

Driver, R. Guesne, E. and Tiberghien, A., (eds.) (1985) *Children's Ideas in Science.* Milton Keynes: Open University Press.

Elstgeest, J. (1985a) 'Encounter, interaction, dialogue'. In *Primary Science: Taking the Plunge.* Ed. Harlen W., London: Heinemann Educational Books.

Elstgeest, J. (1985) 'The right question at the right time'. In *Primary Science: Taking the Plunge.* Ed. Harlen W., London: Heinemann Educational Books.

Elstgeest, J. and Harlen, W. (1990) *Environmental Science in the Primary Curriculum.* London, Paul Chapman Publishing.

Emery H., Saunders, N., Dann, R. and Murphy, R. (1990) *Topics in Assessments 6: Record Keeping.* London: Longman.

Evans, K.M. (1965) *Attitudes and Interests in Education.* London: Routledge and Kegan Paul.

Faire, J. and Cosgrove, M. (1988) *Teaching Primary Science.* Waikato Education Centre, Hamilton, New Zealand

Galton, M. J. Simon, B. and Croll, P. (1980) *Inside the Primary Classroom.* London: Routledge and Kegan Paul.

Gipps, C. (1990) *Assessment: A Teacher's Guide to the Issues.* London: Hodder and Stoughton.

Glover, J. (1985) Case Study 1. Science and project work in the infant school. In Open University EP531: *Primary Science – Why and How?* Block 1 Study Book. Milton Keynes: Open University Press.

Gunstone, R. and Watts, M. (1985) 'Force and Motion'. In *Children's Ideas in Science*, eds. Driver, R. Guesne, E. and Tiberghien, A. Milton Keynes: Open University Press.

Harlen, W. (1985) *Teaching and Learning Primary Science.* London: Paul Chapman Publishing.

Harlen, W. (ed.) (1985) *Primary Science: Taking the Plunge.* London: Heinemann Educational Books.

Harlen, W., Darwin, A. and Murphy, M. (1977) *Match and Mismatch: Raising Questions.* Edinburgh: Oliver and Boyd.

Harlen, W. and Jelly, S. (1989) *Developing Science in the Primary Classroom.* Edinburgh: Oliver and Boyd.

Harlen, W., Macro, C., Schilling, M., Malvern, D. Reed, K. (1990) *Progress in Primary Science.* London: Routledge.

Hawking, S.W. (1988) *A Brief History of Time.* London: Bantam Press.

Holmes, P. (1990) Editorial, *Primary Science Review*, No. 15 p. 2.

Hopkins, D. (1985) *A Teacher's Guide to Classroom Research.* Milton Keynes: Open University Press.

Howe, C. (1990) 'Grouping children for effective learning in science', *Primary Science Review*, No 13, p. 26–7.

IPSE (1988) *The School in Focus.* Hatfield: ASE.

Jameson, S. and Adams, T. (1989) 'A rising star of hope' *Primary Science Review*, Summer, p. 9.

Jelly, S. J. (1985) 'Helping children to raise questions – and answering them'. In *Primary Science: Taking the Plunge.* Ed. Harlen W. London: Heinemann Educational Books.

Kruger, C., and Summers, M., (1989) 'An investigation of some primary teachers' understanding of changes in materials.' *School Science Review*, 71 (255), pp. 17–30.

Layton, D. (1990) *Inarticulate Science.* Occasional Paper No 17. Department of Education, University of Liverpool.

Meadows, J. (1988) 'Grass and parachutes: science and data processing'. *Primary Science Review*, No. 8, p. 4.

National Curriculum Council (NCC) (1989) *Science: Non-Statutory Guidance* York: NCC.

Northern Ireland Schools Examinations and Assessment Council (NISEAC) 1990 *Pupil Assessment In Northern Ireland Arrangements at Key Stages 1, 2, and 3.* Belfast: NISEAC.

Ovens, P. (1987) 'Ice balloons' *Primary Science Review*, No 3 pp. 5–6.

Palmer, D. (1988) 'Monitoring children's progress' *Primary Science Review*, No 7, pp. 26–7.

Paterson, V. (1987) 'What might be learnt from children's writing in primary science?' *Primary Science Review* No. 4, pp. 17–20.

Popper, K., (1988) 'Science: Conjectures and Refutations'. In *Introductory Reading in the Philosophy of Science* ed. E.D.

226

Klemke et al. New York: Prometheus Books.
Russell, T., and Harlen, W. (1990) *Assessing Science in the Primary Classroom: Practical Tasks*. London: Paul Chapman Publishing.
Russell, T. Longden, K. and McGuigan, L. (1991) *Primary SPACE project Report: Materials*. Liverpool University Press.
Russell, T. and Watt, D. (1990) *Primary SPACE project Report: Growth*. Liverpool University Press.
Schilling, M., Hargreaves, L., Harlen, W. with Russell, T. (1990) *Assessing Science in the Primary Classroom: Written Tasks*. London: Paul Chapman Publishing.
Schools Examinations and Assessment Council (SEAC) 1991 *Recorder*, No. 7.
Science Horizons (1982) West Sussex 5–14 Scheme. London: Macmillan Education.
Scottish Office Education Department (SOED) (1990) *Curriculum and Assessment in Scotland: A policy for the '90s*. Working Paper No 1. Edinburgh SOED.
SPACE Project Teachers' Handbook. Forthcoming. London: Collins.
Stewart, J. (1985) *Exploring Primary Science and Technology with Microcomputers*. London: Council for Educational Technology.
Waterman, A. (1990) 'A topic approach to the National Curriculum' *Primary Science Review*, No. 15, pp. 19–21.
Watt, D. and Russell, T. (1990) *Primary SPACE Project Report: Sound* Liverpool University Press.
Wilson, R. J. and Rees, R. (1990) 'The ecology of assessment: evaluation in educational settings, *Canadian Journal of Education*, Vol. 15 pt. 3, pp. 215–228.
Ziman, J. (1968) *Public Knowledge*. Cambridge University Press.

Index